Springer Theses

Recognizing Outstanding Ph.D. Research

Aims and Scope

The series "Springer Theses" brings together a selection of the very best Ph.D. theses from around the world and across the physical sciences. Nominated and endorsed by two recognized specialists, each published volume has been selected for its scientific excellence and the high impact of its contents for the pertinent field of research. For greater accessibility to non-specialists, the published versions include an extended introduction, as well as a foreword by the student's supervisor explaining the special relevance of the work for the field. As a whole, the series will provide a valuable resource both for newcomers to the research fields described, and for other scientists seeking detailed background information on special questions. Finally, it provides an accredited documentation of the valuable contributions made by today's younger generation of scientists.

Theses may be nominated for publication in this series by heads of department at internationally leading universities or institutes and should fulfill all of the following criteria

- They must be written in good English.
- The topic should fall within the confines of Chemistry, Physics, Earth Sciences, Engineering and related interdisciplinary fields such as Materials, Nanoscience, Chemical Engineering, Complex Systems and Biophysics.
- The work reported in the thesis must represent a significant scientific advance.
- If the thesis includes previously published material, permission to reproduce this must be gained from the respective copyright holder (a maximum 30% of the thesis should be a verbatim reproduction from the author's previous publications).
- They must have been examined and passed during the 12 months prior to nomination.
- Each thesis should include a foreword by the supervisor outlining the significance of its content.
- The theses should have a clearly defined structure including an introduction accessible to new PhD students and scientists not expert in the relevant field.

Indexed by zbMATH.

P. Elli Stamatopoulou

Strong Light–Matter Interactions in Extreme Plasmonic and Mie-Resonant Systems

Doctoral Thesis accepted by
the University of Southern Denmark, Odense, Denmark

Springer

Author
Dr. P. Elli Stamatopoulou
Institute of Nanotechnology
Karlsruhe Institute of Technology
Karlsruhe, Germany

Supervisors
Assoc. Prof. Christos Tserkezis
POLIMA—Center for Polariton-Driven
Light-Matter Interactions
University of Southern Denmark
Odense, Denmark

Prof. N. Asger Mortensen
POLIMA—Center for Polariton-Driven
Light-Matter Interactions
University of Southern Denmark
Odense, Denmark

Asst. Prof. Christian Wolff
POLIMA—Center for Polariton-Driven
Light-Matter Interactions
University of Southern Denmark
Odense, Denmark

ISSN 2190-5053 ISSN 2190-5061 (electronic)
Springer Theses
ISBN 978-3-031-67906-3 ISBN 978-3-031-67907-0 (eBook)
https://doi.org/10.1007/978-3-031-67907-0

This Springer imprint is published by the registered company Springer Nature Switzerland AG
The registered company address is: Gewerbestrasse 11, 6330 Cham, Switzerland

Στην οικογένειά μου και τον Κάρλος

Supervisors' Foreword

Humans understand the world primarily through vision. Light comes from a natural or artificial source, it illuminates the piece of matter that we are interested in, and our eyes collect the light reflected or scattered by this object and send the signal to our brains to process it. This makes light one of the most fundamental physical concepts in our everyday lives; understanding its origin and nature has been a major philosophical and scientific quest for centuries. It is thus somehow a surprise that even today, more than 150 years after Maxwell formulated the accepted theory of electromagnetic phenomena, unexpected effects can emerge by studying light. Yet these surprises typically originate not from light itself, but rather through its interaction with matter. As technology improves, new paradigms emerge, especially when matter itself is of sizes comparable to its wavelength, and aspects of the dual wave–particle nature of light become relevant. This is why the fields of nanophotonics and nano-optics, which lie at the intersection where electrical engineering, solid-state physics, materials science, and chemistry meet, keep surprising the corresponding scientific communities at such a steady pace.

The thesis of P. Elli Stamatopoulou deals with the microscopic description of light–matter interactions in a series of different incarnations relevant to active research in nanophotonics. Elli explores light–matter coupling in extreme situations where the size of the particles calls for corrections in the traditional theoretical description, and she considers all three basic light sources in nanophotonics: plane waves, dipoles, and electron beams. She starts with the description of the optical response of extremely small metallic nanoparticles, for which the wavelength of light is comparable to other intrinsic characteristic lengths such as the Fermi wavelength, and implements corrections that account for effects of quantum origin. In these situations, she shows how the coupling of nearby quantum emitters to plasmons is modified by these corrections, both in the weak-coupling regime (molecular emission and fluorescence modified by the presence of a plasmonic system), and in the strong-coupling regime (formation of hybrid half-light–half-matter states known as polaritons). She then performs a similar exploration for the case of larger, yet still subwavelength dielectric particles, which showcase resonances of electric and magnetic multipolar character, each with its own strengths in tailoring light–matter interactions. And, to add yet another dimension, the thesis also considers chiral systems, objects of large biological interest that cannot be superimposed onto their mirror image, to introduce and explore the concept of chiral polaritons. To better visualize the processes that take place in such systems, the final chapter of the thesis departs from plane wave or dipole excitation and delves into the field of electron-beam spectroscopy, which introduces better spatial and temporal resolution in nanophotonic activities. Here, Elli builds on existing literature to present an analytical treatment of an electron beam either passing nearby or traversing a spherical particle and explains how to interpret cathodoluminescence and electron energy-loss spectra.

In all of these cases, Elli presents or develops for the first time analytical expressions suitable for the description of light–matter interactions in an efficient, computationally affordable way that further allows for the development of physical understanding and intuition. All theory is presented with exceptional clarity, while still managing to be concise and avoiding superfluous deviations. The presentation and analysis of all concepts and results suggest a significant level of understanding by the author, and they allow a newcomer to quickly familiarize with the underlying physics. At the same time, the technical details included in the main text and the appendices allow one to first reproduce the results of this thesis and then embark into their own explorations of light–matter interactions. We are thus extremely happy that we assisted, even a little bit, in the creation of a thesis with such high educational value.

Odense, Denmark
January 2024

Assoc. Prof. Christos Tserkezis
Prof. N. Asger Mortensen
Asst. Prof. Christian Wolff

Abstract

In this thesis, we develop and apply analytic methods for analyzing the interaction of light and matter in extreme plasmonic and Mie-resonant systems. Over the past few decades, optical excitations in matter, either in the form of plasmons in metals or Mie resonances in high-index dielectrics, have emerged as potential building blocks for novel photonic devices that require efficient light manipulation. Owing to their ability to strongly enhance and confine incident electromagnetic fields, plasmonic elements are widely employed in sensors, nanoantennas, or metasurfaces, for applications ranging from optical and quantum communications to (bio)sensing and polaritonic chemistry. In applications where the high Ohmic losses and the subsequent heat generation are detrimental, dielectrics are being explored as an alternative for low-loss resonators. Even though Mie-resonant systems do not achieve as high field confinement as plasmonic ones typically do, they support modes of both electric and magnetic multipolar character, thus exhibiting a richer optical response. Through multipolar interference, dielectrics yield high-quality factors and controllable radiation emission, making them eligible components for advanced nanolasers, topological photonics, and sensing.

The thesis addresses the optical excitations sustained in metallic and dielectric nanospheres and their interaction with light, quantum emitters, and fast electrons. First, we discuss the strong coupling and the resulting hybrid light–matter modes that form when certain configurations of core–shell nanoparticles are illuminated. The structures studied here consist of an open nanocavity, in the form of one or more nanoparticles supporting plasmonic or Mie modes, encapsulated in a shell that sustains excitons, i.e. electronic transitions such as the ones found in low-dimensional semiconductors or organic molecules. We then proceed to investigate the impact of this modal hybridization on the chiroptical response of chiral nanostructures and find that circular dichroism can act as an imprint of strong coupling. Secondly, we examine the weak coupling of quantum emitters featuring a single optical transition with metallic or dielectric nanocavities and compare how the two types of materials modify the excitation and emission properties of the emitter when the system is excited by light. Following the direction of modern photonics towards the integration of ever smaller components, in both coupling regimes, we explore systems of nanometric dimensions, where the limits of classical and local material-response descriptions are inevitably exposed. Rather than developing a radically new theory, we show how nonclassical effects in plasmonics, such as electron-density spill-out, screening, and enhanced Landau damping, can be incorporated into classical electrodynamics via nonlocal corrections, which apply mainly to the boundary conditions.

However, to study the properties of extreme-nanoscale structures, spectroscopy techniques of atomic resolution are required. To this end, we redirect our focus from light to electrons as sources for probing optical excitations in matter and develop analytic tools for simulating and interpreting measurements in cathodoluminescence and electron energy-loss spectroscopy. We cast particular attention on metallic and dielectric nanospheres excited by penetrating electron beams and examine the coexistence of the resonant modes of the structure together with other potential emission or electron energy-loss mechanisms. In this regard, we provide a detailed account of the interplay between Mie resonances and transition radiation in the cathodoluminescence spectra of silicon nanoparticles.

Acknowledgements

First and foremost, I would like to express my deepest gratitude to my supervisors Assoc. Prof. Christos Tserkezis, Asst. Prof. Christian Wolff, and Prof. N. Asger Mortensen for their guidance and support throughout my Ph.D. studies. I am deeply indebted to my principal advisor Christos, whose mentorship has been truly invaluable during my academic journey. You have supported me in every step of the process, with your scientific expertise, your patience and encouragement, and, most importantly, your culinary skills. I could not have asked for a better supervisor. I would like to extend my appreciation to Christian, whose brilliance has been an incredible source of inspiration, knowledge, and witty puns. I am profoundly grateful to Asger, who has been an exemplary scientist and group leader. Much like a skilled winemaker, you have masterfully guided my studies, from the initial crushing of ideas to the insightful evaluation of the final product, while providing an excellent environment where my ideas could ferment and age into a rich body of work.

I extend my thanks to my colleagues in POLIMA and CNO, whose presence has been instrumental in shaping the ideas presented in this thesis, and my sugar tolerance. I especially thank Mikkel, Nikos, Pedro, Theis, Martin, and Assoc. Prof. Joel Cox for proofreading parts of the thesis, and Louise for her support in handling administrative tasks.

I would also like to thank Prof. Javier Aizpurua and Dr. Rubén Esteban for their hospitality at the Materials Physics Center in San Sebastián during my Ph.D. external research stay. The interaction with the group has significantly broadened my understanding of the concepts explored in this thesis. Special appreciation goes to Dr. Mattin Urbieta for his generous sharing of expertise in the Boundary Element Method.

I wish to express my gratitude to the members of my Ph.D. thesis committee, Prof. Päivi Törmä, Prof. Javier García de Abajo, Assoc. Prof. Antonio Fernández-Domínguez and Assoc. Prof. Jonas Beermann for agreeing to take time out of their busy schedules to serve on my thesis assessment committee.

On a personal level, I wish to acknowledge the people who stood by my side through trials and triumphs. I would like to thank my family for their unwavering support and their steadfast supply of extra virgin olive oil. My most heartfelt thanks to

Carlos for his unconditional love and encouragement, and for his impromptu decision to include me as a co-author in all his future papers.

Finally, I acknowledge the financial support provided by VILLUM FONDEN (Grant Number 16498), which funded CNO, and the Danish National Research Foundation (Project Number. DNRF165), which is funding POLIMA. I also acknowledge the Zonta Foundation for Women for honoring me with the Zonta Ph.D. scholarship for women in natural sciences and technology (September 2021).

Contents

Abbreviations

BP(P)	Bulk Plasmon (Polariton)
CD	Circular Dichroism
CED	Classical Electrodynamics
CHO	Coupled Harmonic Oscillators
CL	Cathodoluminescence
CR	Cherenkov Radiation
DGTD	Discontinuous Galerkin Time-Domain
DS	Differential Scattering
ED (MD)	Electric (Magnetic) Dipole
EEL(S)	Electron Energy Loss (Spectroscopy)
EM	Electromagnetic
EO (MO)	Electric (Magnetic) Octupole
EQ (MQ)	Electric (Magnetic) Quadrupole
FEG	Free Electron Gas
FEM	Finite-Element Method
GNOR	Generalized Nonlocal Optical Response
GR	Geometric Resonance
HDM	Hydrodynamic Drude Model
IT	Interband Transition
LCP (RCP)	Left (Right) Circularly Polarized
LRA	Local Response Approximation
LSP	Localized Surface Plasmon
NP	Nanoparticle
OR	Optical Rotation
QE	Quantum Emitter
QED	Quantum Electrodynamics
SDB	Size-Dependent Broadening
SEF	Surface-Enhanced Fluorescence
SEM	Scanning Electron Microscope
SERS	Surface-Enhanced Raman Scattering
SP(P)	Surface Plasmon (Polariton)

SRF	Surface Response Formalism
SRM	Specular-Reflection Model
TDDFT	Time-Dependent Density Functional Theory
TE (TM)	Transverse Electric (Magnetic)
TEM	Transmission Electron Microscope
TMDC	Transition-Metal Dichalcogenide
TR	Transition Radiation
UEM	Ultrafast Electron Microscopy
UP (LP)	Upper (Lower) Polariton

Symbols

$\mathbf{A}$	Vector or matrix
A	Scalar or norm of real-valued vector $\mathbf{A}$
$\hat{\mathbf{a}}$	Unitary vector
$\overleftrightarrow{a}$	Tensor
n	Refractive index
λ	Wavelength
$k(\mathbf{k})$	Wavenumber (wavevector)
d	Distance
$\mathbf{r}$	Position vector
t	Time
ω	(Angular) frequency
$\mathbf{E}$	Electric field
$\mathbf{D}$	Electric displacement field
$\mathbf{B}$	Magnetic field/induction/flux density
$\mathbf{H}$	Magnetic/magnetization field
ρ	Volume charge density
$\mathbf{J}$	Volume current density
$\mathbf{F}$	Force
$(e)q$	(Electron) charge
$\mathbf{v}$	Velocity
$\mathbf{P}$	Polarization density
$\mathbf{M}$	Magnetization density
$\mathbf{p}$	Electric dipole moment or momentum
χ_e	Electric susceptibility
ε	Relative permittivity/dielectric function
ξ, ζ	Magnetoelectric coupling parameters
μ	Relative permeability
c	Speed of light in vacuum
σ	Surface charge density or conductivity or optical cross section
$\mathbf{K}$	Surface current density
$\mathbf{A}$	Vector potential

φ	Scalar potential
χ	Tellegen parameter
κ	Chirality/Pasteur parameter
C	Optical chirality density
τ	Relaxation or decay time
γ	(Damping or excitation or decay) rate
n	Electron density
m_e	Electron mass
E	Energy
f	Oscillator strength
$\hbar$	Reduced Planck constant
$\mathscr{p}$	Hydrodynamic pressure
$\mathscr{b}$	Hydrodynamic parameter
ℓ, m	Orbital angular momentum quantum numbers
Z	Impedance
α	Polarizability
g	Coupling strength
η	(Excitation or fluorescence) enhancement
W	Work
P	Power
$\mathscr{Q}$	Quantum yield
$\xi_{\rm NL}$	Nonlocal length
$\mathscr{D}$	Diffusion constant
$\mathscr{n}$	Convective-diffusive parameter
$d_{\perp/\parallel}$	Feibelman d-parameters
$\boldsymbol{\pi}$	Surface polarization density
β, γ	Lorentz kinematic factors
b	Impact parameter
q	Transverse momentum
$\Gamma_{\rm CL/EEL/CR}$	(CL/EEL/CR) probability
$\phi_{\rm out}$	Microscope collection angle
σ_e	Beam width

Chapter 1
Introduction

Abstract This chapter introduces the field of nano optics. Emphasizing the historical significance of understanding and controlling light, we discuss the key discoveries that consolidated the modern theories of light, matter, and the interaction thereof, and motivated the passage from classical to nano optics. The chapter outlines the main topics addressed in this thesis, placing them in context within the broader scope of the field.

1.1 From Optics to Nano Optics

Unlocking and controlling the properties of light has always been a major interest of humankind. Polished bronze disks were used by the ancient Egyptians as mirrors, as early as 2900 BC. The first reported attempts to actively manipulate light date back to the 8th century BC in Mesopotamia, where Assyrian craftsmen designed the first optical lenses [1]. Advanced ancient civilizations quickly realized the benefits of understanding the principles of optics, and invented lenses and mirrors for diverse uses ranging from simple magnifying glasses to alleged warfare devices. This realization, combined with a strong curiosity to understand the nature of light, gave birth to the field of classical optics. Two key discoveries consolidated the modern perception of light: Maxwell's description of it as a propagating electromagnetic (EM) wave (1861–1862) [2, 3], and Einstein's discovery of the photoelectric effect (1905) [4], positing that light travels in bundles of energy named *photons*, thus revealing its quantum nature. It became clear, at that moment, that light can be best understood through its interaction with matter, and conversely.

From the naked eye, to magnifying lenses and compound microscopes, light has been our oldest and most reliable tool to study matter. Since the 1600s and for centuries, the optical properties of matter were studied via optical microscopes. Over the following years a large number of elaborate designs was developed to improve the spatial resolution, until Ernst Abbe demonstrated its fundamental limit: light of wavelength λ traveling through a medium with refractive index n and converging to a spot with half-angle θ has a minimum resolvable distance [5]

P. E. Stamatopoulou, *Strong Light-Matter Interactions in Extreme Plasmonic and Mie-Resonant Systems*, Springer Theses, https://doi.org/10.1007/978-3-031-67907-0_1

$$d = \frac{\lambda}{2n \sin\theta} . \tag{1.1}$$

Abbe's diffraction limit is a consequence of the fact that the large k (wavenumber) components of light scattered by an object decay exponentially away from it, and thus cannot be observed in the far field. This means that a conventional optical microscope of numerical aperture ($\text{NA} = n \sin\theta$) close to 1 operating in the visible wavelength range 380–750 nm, cannot resolve details smaller than 200 nm. Unfortunately, viruses (100 nm), proteins (10 nm) and other less complex biomolecules of interest (1 nm) remain beyond the reach of conventional optics. In other words, the diffraction limit of light prevents optics from going nano. Physicists have come up with a number of elaborate schemes to successfully push the limits of optics to the nanoscale. Among those, and at the intersection between material science and electromagnetism, lies the field of *plasmonics*.

Since light is an EM wave, it interacts with the charged particles comprising matter. Being highly mobile, the free electrons in metals interact strongly with light; the effects of this interplay are exactly the focus of plasmonics. From a quantum-mechanical point of view, the elementary building blocks partaking are the photons and the *plasmons*. Predicted by Pines and Bohm, plasmons represent the quasiparticles of the collective oscillation of conduction electrons in metals [6]. In 1957 Ritchie postulated that such oscillations can be excited on a metal–air interface, introducing for the first time what was later termed a *surface plasmon* (SP) [7]. When a metallic surface is illuminated, photons and SPs—under the right conditions—couple, resulting in the formation of a *surface plasmon polariton* (SPP) that bears the properties of both light and matter. Apart from the scientific curiosity over the physics underlying this bizarre hybrid, part of the tremendous appeal of SPPs stems from the unique set of properties they feature: strong enhancement and concentration of the incident light in subwavelength volumes [8]. SPP waves propagate along a metal–dielectric interface like ripples on a pond, while remaining tightly confined on it. Plasmonic structures can confine the incident visible light to volumes as small as a few cubic nanometers, and enhance the near-field component of light, which contains the high-k information below the diffraction limit. In the pursuit to minimize photonic devices, which has been a clear trend in photonics right from the beginning, these attributes have consistently motivated the use of plasmons in simple or more sophisticated designs. The resulting platforms have been proposed or already employed for applications ranging from optical communications [9], lasing [10] and photovoltaics [11], to nanomedicine [12] and catalysis [13].

Plasmons offer more than one ways to manipulate light. Due to their high sensitivity to the local optical environment, as well as their ability to influence its properties, metallic structures have proven particularly valuable in the design of sensors [14]. Ever since Purcell's observation (1941) that the emission properties of emitters, such as two-level atoms, are modified in the presence of a structured environment [15], SPs have been employed to improve the efficiency of diverse radiative processes. In surface-enhanced fluorescence (SEF) and surface-enhanced Raman scattering (SERS), the fluorescence and Raman emission of organic molecules, respectively, are enhanced by placing them near metallic surfaces with nanoscale roughness or near metallic nanoparticles (NPs) [16]. Besides what has already been established, there is yet an even larger number of applications envisioned. Nonetheless, in order to be fully and efficiently realized, they require further investigation of the properties of plasmons.

Since SPs in plasmonic systems proved to couple efficiently to emitters formed by electronic transitions in matter, the next natural step was to explore the regime of strong coupling between them. In quantum optics, strong light–matter coupling refers to the reversible energy exchange between a quantum emitter (QE), typically modeled as a two- or three-level system, and a cavity formed by two mirrors enclosing the QE; upon illumination, the light trapped in the cavity excites the QE, which then decays by spontaneously emitting a photon in the cavity, only to be absorbed again, and so on [17]. As a consequence, the properties of the cavity–atom system combine both light and matter characteristics. The emergence of this hybrid state has ignited a drive to extend the concept of strong coupling to the field of photonics and exploit the properties of the resulting polaritons for polaritonic lasing [18], Bose-Einstein condensation [19], and polaritonic chemistry [20]. In the mesoscopic world, resonant NPs that confine EM modes at their surface or in their volume can play the role of an open cavity, and couple to electronic transitions in single atoms or to collections of two-level systems, such as electron–hole pairs (*excitons*) in excitonic media. However, since metals suffer from high Ohmic losses, the plasmon–exciton (*plexciton*) system must be carefully designed to ensure that the energy exchange between the two constituents is faster than their respective dissipation rates. The first efforts to demonstrate this interaction in solid-state systems were reported in the 1990s in microcavities involving inorganic semiconductors, mostly at cryogenic temperatures [21]. A key development to achieve robust strong coupling at room temperature was the introduction of organic molecules [22, 23], such as J–aggregates found in organic dyes, which support collective excitonic transitions of energies at the optical wavelengths. The tight field confinement provided by plasmonic cavities, together with the strong dipole moments of organic molecules, have proven to be an ideal combination for strong-coupling studies [24]. The appeal of all these prospects is in fact so strong, that the nano-optics community usually finds a way to surpass the practical limitations imposed by the high losses of metals [25], or seeks ways to work around them [26].

Despite those attempts, oftentimes the losses completely hinder the desired function of a photonic device, in which case plasmonic nanostructures are ultimately unsuitable. Owing to their low losses, high-index dielectrics have been recently proposed as alternative substitutes, with silicon being the most typical material of choice. The resonant behavior of dielectrics emerges from the collective oscillation of bound electrons, and is characterized by electric- and magnetic-type multipolar modes, visualized through the so-called *Mie resonances* [27]. Dielectric NPs feature properties similar to their metallic counterparts, such as high sensitivity to their optical environment and strong field enhancement. Even though silicon cannot compete with the field-confinement capabilities of metals, several works have demonstrated both theoretically and experimentally that robust strong coupling [28, 29] and an appreciable fluorescence enhancement [30, 31] are possible in the interaction of dielectric nanocavities with QEs. At the same time, unlike nonmagnetic metals, certain dielectric materials can exibit natural optical chirality, meaning that they distinguish between left- and right-handed light, thus enabling the selective manipulation of polarized light. These developments have shifted the focus of the nanophotonics community to all-dielectric designs for photonic applications, also benefiting from the compatibility with existing silicon-based technologies [32, 33].

1.2 Beyond the Local Response Approximation

Whether it concerns applications relying on field enhancement, like SERS, or strong coupling, the quest of nanophotonics has always been to squeeze light in ever smaller volumes, by reducing both the size of the structures involved and their separations. However, this mission soon exposed the insufficiency of the existing theories, which were based on classical electrodynamics (CED) and the local response of materials, in describing light–matter interactions at the nanoscale [34, 35]. This was characteristically highlighted in the study of metallic dimers separated by nanometric gaps, a structure that garnered significant attention due to the fact that the local response approximation (LRA) predicted a huge field enhancement within the formed cavity; in fact, a divergent field enhancement with reducing separation [36]. Even before such nanogaps were actually produced, it had become clear that this classically predicted divergence could not be feasible, leading to the quest for understanding the physical mechanisms that regularize it.

In solid-state physics, it has been known for decades that local dielectric functions prove inadequate for the full description of metals, once the length scales involved become comparable to a characteristic intrinsic length of the system, such as the Fermi wavelength or the electron mean free path in a more classical description [37]. A spatially nonlocal, and therefore wavevector-dependent permittivity is required [38], which, already in its most simple form, accounts for the effect of screening: charges are spatially extended, and inner electrons screen the interaction with the outer ones. Oftentimes, models developed in solid-state physics have been successfully applied in plasmonics to reproduce nonlocal effects, for instance Bloch's hydrodynamic treatment of the electron dynamics [39]. Albeit successful in describing screening in noble metals, the hydrodynamic approach falls short in explaining the experimental observations in the optical response of alkali metals like sodium, where other effects of quantum origin dictate the electron dynamics. Over the years, more sophisticated semiclassical schemes have been developed to encompass such mechanisms and elucidate our experimental observations. The need for better modeling of light and matter in nanometric objects has launched the field of the so-called *mesoscopic* plasmonics [40]; by introducing quantum corrections, while still operating within CED, the mesoscopic approach circumvents the fully quantum mechanical treatment, which comes with a high—in certain occasions, prohibitive—computational cost, thus bridging the gap between classical and quantum electrodynamics (QED).

While currently modern fabrication has provided access to metallic systems of almost atomistic sizes (∼2 nm) [41], this is definitely not the case for dielectrics. Nonlocality in all-dielectric platforms is rarely a subject of investigation—neither will be in this thesis—, not only because of the challenges in fabrication, but most importantly because their main selling point, i.e. the richness of their optical response is only present in structures above 50 nm, where finite-size effects are completely masked by the bulk response of the material. Nonetheless, a two-fluid hydrodynamic description for electrons and holes in semiconducting media has been developed that can potentially capture nonlocal effects in some dielectrics [42].

1.3 Spectroscopy Under the Electron Microscope

Possessing the theoretical tools to analyze light–matter interactions at the nanoscale is, of course, of little value if they remain beyond the reach of our imaging and spectroscopic methods. We already described how SPs are employed in spectroscopy in order to enhance the response of nearby emitters, owing to their ability to enhance near fields where the high-resolution information in encoded. However, when the metallic or dielectric structures are themselves the subject of study, exciting and recording the high-k fields sustained therein requires a localized probe.

Electron microscopy has proven exceptionally successful for the study of optical modes in matter, since the localized nature of electrons can access the near field generated by the collective motion of the electrons [43, 44]. In fact, in Ritchie's original work, as well as in the experimental confirmation by Powell and Swan [45], fast electrons were used to excite the SPs. Despite being an active field for decades, it was only recently that the mapping of plasmon fields was achieved with nanometer resolution [46, 47]. This has massively rekindled the interest on studying the optical excitations in matter under the electron microscope.

With the optimal resolution ultimately being limited by the de Broglie wavelength of the electron (below 0.1 Å at energies above 20 keV), the state-of-the-art electron microscopes nowadays achieve atomic spatial resolution and sub-meV energy resolution. Several electron spectroscopy techniques have been devised, depending on the type of the signal collected, including inelastic losses of the swift electrons in electron energy-loss spectroscopy (EELS) or far-field radiation in cathodoluminescence (CL) spectroscopy. Recent progress in instrumentation has even added femtosecond temporal resolution in EEL and CL spectra, launching the field of ultrafast electron microscopy (UEM) [48, 49], and paving the way for the dynamic study of light–matter interactions [50, 51]. Considering the ongoing progress and future potential of electron-beam spectroscopies, the need for theoretical methods to interpret the plethora of experimental data is now imperative.

1.4 Outline

Motivated by all the efforts and prospects presented through the exploration of light and matter at the nanoscale, this thesis examines the interaction of collective optical excitations in metallic and dielectric NPs with light, emitters, and swift electrons. The current document is an edited version of the official thesis, typeset to the format of *Springer Theses*. The work presented here is related to the publications enlisted on page 177. The thesis is organized in six chapters, and is accompanied by Appendices A–F that contain detailed derivations of most expressions found in the chapters. Below, we provide a short description of the content of each chapter.

Chapter 2 | **Theory**

Chapter 2 begins with a review of the fundamentals of CED and the properties of the materials under study. After establishing the basic principles and analytic methods regarding light–matter interactions in the classical local picture, we next delve into the mesoscopic theory of plasmonics, and review (here and in Publication C) a selection of models that encompass quantum corrections in an otherwise CED framework. In the final section of the chapter, we describe the numerical approach employed in the fifth chapter of the thesis.

Chapter 3 | **Strong Coupling in Photonics**

In Chap. 3 we theoretically explore the properties of spherical core–shell configurations, comprising nanocavities that sustain either plasmon or Mie resonances, strongly coupled to an excitonic shell. We specifically address two topics; firstly, we investigate the influence of finite-size effects in nanometric plexcitonic systems. Secondly, we elaborate on the findings of Publication B, concerning the modification of the chiral response of optically active configurations due to strong coupling to excitons.

Chapter 4 | **Weak Coupling of Nanoparticles with Quantum Emitters**

Chapter 4 examines the weak interaction between spherical nanocavities in close proximity to QEs of varying positions and orientations. We compare the fluorescence enhancement considering dielectric and metallic nanospheres, and discuss the results presented in Publication E. We then analyze the excitation and emission processes of QEs near nanometric plasmonic cavities as predicted by a quantum-corrected approach.

Chapter 5 | **Interaction of Nanoparticles with Electron Beams**

In Chap. 5 we present the analytic electrodynamic and relativistic theory developed in Publication A, pertaining to spherical NPs interacting with swift electrons passing by or through their volume. The method is applied to study the response of metallic nanospheres, and compared to results obtained with a state-of-the-art numerical method for problems in CED, namely the discontinuous Galerkin time-domain (DGTD) method. The final section of the chapter shows experimental and theoretical CL spectra of silicon NPs, published in Publication D, and addresses the interplay between the emission originating from the modes of the structure with other sources of coherent radiation.

Chapter 6 | Summary and Conclusions

Chapter 6 summarizes the thesis and presents the main conclusions.

References

1. Lindberg DC (1976) Theories of vision from al-Kindi to Kepler. University of Chicago Press
2. Maxwell JC (1861) Philos Mag 21:338. https://doi.org/10.1080/14786446108643067
3. Maxwell JC (1862) Philos Mag 23:12. https://doi.org/10.1080/14786446208643207
4. Einstein A (1905) Ann Phys 4:132. https://doi.org/10.1002/andp.19053220607
5. Born M, Wolf E (1999) Principles of optics, 7th ed. Cambridge University Press
6. Pines D, Bohm D (1952) Phys Rev 85:338. https://doi.org/10.1103/PhysRev.85.338
7. Ritchie RH (1957) Phys Rev 106:874. https://doi.org/10.1103/PhysRev.106.874
8. Schuller JA, Barnard ES, Cai W, Jun YC, White JS, Brongersma ML (2010) Nat Mater 9:193. https://doi.org/10.1038/nmat2630
9. Ozbay E (2006) Science 311:189. https://doi.org/10.1126/science.1114849
10. Hakala TK, Rekola HT, Väkeväinen AI, Martikainen J-P, Nečada M, Moilanen AJ, Törmä P (2017) Nat Commun 8:13687. https://doi.org/10.1038/ncomms13687
11. Atwater HA, Polman A (2010) Nat Mater 9:205. https://doi.org/10.1038/nmat2629
12. Jain PK, Huang X, El-Sayed IH, El-Sayed MA (2007) Plasmonics 2:107. https://doi.org/10.1007/s11468-007-9031-1
13. Christopher P, Xin H, Linic S (2011) Nat Chem 3:467. https://doi.org/10.1038/nchem.1032
14. Willets KA, Van Duyne RP (2007) Annu Rev Phys Chem 58:267. https://doi.org/10.1146/annurev.physchem.58.032806.104607
15. Purcell EM, Torrey HC, Pound RV (1946) Phys Rev 69:37. https://doi.org/10.1103/PhysRev.69.37
16. Giannini V, Fernández-Domínguez AI, Heck SC, Maier SA (2011) Chem Rev 111:3888. https://doi.org/10.1021/cr1002672
17. Törmä P, Barnes WL (2015) Rep Prog Phys 78:013901. https://doi.org/10.1088/0034-4885/78/1/013901
18. Ramezani M, Halpin A, Fernández-Domínguez AI, Feist J, Rodriguez SR-K, Garcia-Vidal FJ, Gómez Rivas J (2017) Optica 4:31. https://doi.org/10.1364/OPTICA.4.000031
19. Kasprzak J, Richard M, Kundermann S, Baas A, Jeambrun P, Keeling JMJ, Marchetti FM, Szymańska MH, André R, Staehli JL, Savona V, Littlewood PB, Deveaud B, Dang LS (2006) Nature 443:409. https://doi.org/10.1038/nature05131
20. Feist J, Galego J, Garcia-Vidal FJ (2018) ACS Photon 5:205. https://doi.org/10.1021/acsphotonics.7b00680
21. Skolnick MS, Fisher TA, Whittaker DM (1998) Semicond Sci Technol 13:645. https://doi.org/10.1088/0268-1242/13/7/003
22. Pockrand I, Brillante A, Möbius D (1982) J Chem Phys 77:6289. https://doi.org/10.1063/1.443834
23. Bellessa J, Bonnand C, Plenet JC, Mugnier J (2004) Phys Rev Lett 93:036404. https://doi.org/10.1103/PhysRevLett.93.036404
24. Fofang NT, Grady NK, Fan Z, Govorov AO, Halas NJ (2011) Nano Lett 11:1556. https://doi.org/10.1021/nl104352j
25. Khurgin JB (2015) Nat Nanotech 10:2. https://doi.org/10.1038/nnano.2014.310
26. Boltasseva A, Atwater HA (2011) Science 331:290. https://doi.org/10.1126/science.1198258

27. García-Etxarri A, Gómez-Medina R, Froufe-Pérez LS, López C, Chantada L, Scheffold F, Aizpurua J, Nieto-Vesperinas M, Sáenz JJ (2011) Opt Express 19:4815. https://doi.org/10.1364/OE.19.004815
28. Tserkezis C, Gonçalves PAD, Wolff C, Todisco F, Busch K, Mortensen NA (2018) Phys Rev B 98:155439. https://doi.org/10.1103/PhysRevB.98.155439
29. Wang H, Ke Y, Xu N, Zhan R, Zheng Z, Wen J, Yan J, Liu P, Chen J, She J, Zhang Y, Liu F, Chen H, Deng S (2016) Nano Lett 16:6886. https://doi.org/10.1021/acs.nanolett.6b02759
30. Albella P, Poyli MA, Schmidt MK, Maier SA, Moreno F, Sáenz JJ, Aizpurua J (2013) J Phys Chem C 117:13573. https://doi.org/10.1021/jp4027018
31. Krasnok A, Glybovski S, Petrov M, Makarov S, Savelev R, Belov P, Simovski C, Kivshar Y (2016) Appl Phys Lett 108:211105. https://doi.org/10.1063/1.4952740
32. Baranov DG, Zuev DA, Lepeshov SI, Kotov OV, Krasnok AE, Evlyukhin AB, Chichkov BN (2017) Optica 4:814. https://doi.org/10.1364/OPTICA.4.000814
33. Staude I, Pertsch T, Kivshar YS (2019) ACS Photon 6:802. https://doi.org/10.1021/acsphotonics.8b01326
34. Doremus RH (1964) J Chem Phys 40:2389. https://doi.org/10.1063/1.1725519
35. Kreibig U, von Fragstein C (1969) Z Phys 224:307. https://doi.org/10.1007/BF01393059
36. Ciracì C, Hill RT, Mock JJ, Urzhumov Y, Fernández-Domínguez AI, Maier SA, Pendry JB, Chilkoti A, Smith DR (2012) Science 337:1072. https://doi.org/10.1126/science.1224823
37. Ashcroft NW, Mermin ND (1976) Solid state physics. Holt-Saunders
38. Lindhard J (1954) Dan Mat Fys Medd 28:1
39. Bloch F (1933) Z Phys 81:363. https://doi.org/10.1007/bf01344553
40. Mortensen NA (2021) Nanophotonics 10:2563. https://doi.org/10.1515/nanoph-2021-0156
41. Scholl JA, Koh AL, Dionne JA (2012) Nature 483:421. https://doi.org/10.1038/nature10904
42. Maack JR, Mortensen NA, Wubs M (2018) Phys Rev B 97:115415. https://doi.org/10.1103/PhysRevB.97.115415
43. García de Abajo FJ (2010) Rev Mod Phys 82:209. https://doi.org/10.1103/RevModPhys.82.209
44. Polman A, Kociak M, García de Abajo FJ (2019) Nat Mater 18:1158. https://doi.org/10.1038/s41563-019-0409-1
45. Powell CJ, Swan JB (1959) Phys Rev 115:869. https://doi.org/10.1103/PhysRev.115.869
46. Bosman M, Keast VJ, Watanabe M, Maaroof AI, Cortie MB (2007) Nanotechnology 18:165505. https://doi.org/10.1088/0957-4484/18/16/165505
47. Nelayah J, Kociak M, Stéphan O, García de Abajo FJ, Tencé M, Henrard L, Taverna D, Pastoriza-Santos I, Liz-Marzán LM, Colliex C (2007) Nat Phys 3:348. https://doi.org/10.1038/nphys575
48. Barwick B, Park HS, Kwon O-H, Baskin JS, Zewail AH (2008) Science 322:1227. https://doi.org/10.1126/science.1164000
49. Barwick B, Flannigan DJ, Zewail AH (2009) Nature 462:902. https://doi.org/10.1038/nature08662
50. Wang K, Dahan R, Shentcis M, Kauffmann Y, Ben Hayun A, Reinhardt O, Tsesses S, Kaminer I (2020) Nature 582:50. https://doi.org/10.1038/s41586-020-2321-x
51. Kfir O, Lourenço-Martins H, Storeck G, Sivis M, Harvey TR, Kippenberg TJ, Feist A, Ropers C (2020) Nature 582:46. https://doi.org/10.1038/s41586-020-2320-y

Chapter 2
Theory

Abstract This chapter outlines the basic concepts and methods of nano optics, laying the theoretical groundwork for the chapters to follow. We begin with classical electrodynamics, describing light through Maxwell's equations and matter within a macroscopic framework. We introduce the fundamentals of light—matter interactions and polaritonics involving metallic or dielectric nanocavities, including Mie theory, classical and mesoscopic plasmonics. Finally, we discuss light—matter interactions under the scope of strong and weak coupling of polaritons with electronic transitions in matter. We conclude this introduction with a brief outline of the numerical method used in the fifth chapter of the thesis.

2.1 Fundamentals of Classical Electrodynamics

2.1.1 Maxwell's Equations

The fundamental physical quantities of an EM problem are the electric field **E** and the magnetic field **B**.[1] In vacuum, and in the absence of free charges and currents, the spatiotemporal dependence $(\mathbf{r}, t)$ of the two fields is described through Maxwell's equations, which in SI units read [1]

$$\nabla \cdot \mathbf{E} = 0\,, \tag{2.1a}$$

$$\nabla \times \mathbf{B} = \varepsilon_0 \mu_0 \frac{\partial \mathbf{E}}{\partial t}\,, \tag{2.1b}$$

$$\nabla \times \mathbf{E} = -\frac{\partial \mathbf{B}}{\partial t}\,, \tag{2.1c}$$

$$\nabla \cdot \mathbf{B} = 0\,, \tag{2.1d}$$

---

[1] The magnetic field **B** is also encountered as magnetic induction or magnetic flux density in literature.

P. E. Stamatopoulou, *Strong Light-Matter Interactions in Extreme Plasmonic and Mie-Resonant Systems*, Springer Theses, https://doi.org/10.1007/978-3-031-67907-0_2

where ε_0 is the vacuum permittivity and μ_0 the vacuum permeability. Equations (2.1b) and (2.1c) reveal that time-varying electric fields give rise to magnetic fields and vice versa. The two fields are connected and we thus speak of *electromagnetic fields*.

In macroscopic media, and in the presence of a free charge, ρ_f, and current, $\mathbf{J}_f$, density that we control externally, one has to include the external sources and the material response to the prescription. An external electric field **E**, applied to a medium viewed as a collection of atoms, displaces the electrons from the nuclei, creating small electric dipoles, each characterized by dipole moment **p**. The dipoles align with the external field, leading to a macroscopic polarization density **P** (electric dipole moment per unit volume). The electric displacement field **D** accounts for both the effect of the electric field applied to the material and the polarization of the medium in response to that external field. Similarly, an external magnetic field leads to a magnetization density **M** (magnetic dipole moment per unit volume), while a time variation of the polarization creates an additional flow of bound charges, which contributes to the total current **J** through a polarization current. The total magnetic field **B** encompasses both external magnetic fields and the effects of magnetization within the material, while the magnetic field created by external sources (ρ_f and $\mathbf{J}_f$) is represented by the magnetic (or magnetizing) **H**-field.[2] Introducing the auxiliary fields **D** and **H**, we can formulate the *macroscopic* Maxwell equations, that depend on the external sources only.

Box 2.1 | Macroscopic Maxwell equations in matter

Gauss's law
$$\nabla \cdot \mathbf{D} = \rho_f \, , \tag{2.2a}$$

Ampère-Maxwell's law
$$\nabla \times \mathbf{H} = \frac{\partial \mathbf{D}}{\partial t} + \mathbf{J}_f \, , \tag{2.2b}$$

Faraday's law
$$\nabla \times \mathbf{E} = -\frac{\partial \mathbf{B}}{\partial t} \, , \tag{2.2c}$$

Absence of magnetic monopoles
$$\nabla \cdot \mathbf{B} = 0 \, . \tag{2.2d}$$

The first two relations can be combined to derive the continuity equation connecting the free charge and current density

$$\frac{\partial \rho_f}{\partial t} + \nabla \cdot \mathbf{J}_f = 0 \, . \tag{2.3}$$

A point charge q traveling with velocity **v** inside an EM field experiences the Lorentz force

$$\mathbf{F} = q(\mathbf{E} + \mathbf{v} \times \mathbf{B}) \, . \tag{2.4}$$

[2] Throughout the thesis we refer to both **B** and **H** as the magnetic field. However, we emphasize that **B** is the external field and **H** the response to it.

Maxwell's equations (2.2), together with the Lorentz force and Newton's second law provide a complete description of the classical dynamics of charged particles interacting with EM fields.

2.1.2 Constitutive Relations

The set of Eqs. (2.2) can be solved once we introduce the relations connecting the auxiliary fields **D** and **H** to **E** and **B**, which are known as *constitutive relations*. In the scope of this study, we consider external fields that are weak enough to assure the *linear* response of the material. In this regard, the external field induces a proportional polarization density[3]

$$\mathbf{P} = \varepsilon_0 \overleftrightarrow{\chi}_e \mathbf{E} \,, \tag{2.5}$$

where $\overleftrightarrow{\chi}_e$ is the electric susceptibility, with symbol ($\leftrightarrow$) denoting a tensor form. Then the displacement field is also proportional to the external field,

$$\mathbf{D} = \mathbf{P} + \varepsilon_0 \mathbf{E} = \overleftrightarrow{\varepsilon} \varepsilon_0 \mathbf{E} \,, \tag{2.6}$$

where the proportionality constant, namely the permittivity $\overleftrightarrow{\varepsilon}\varepsilon_0 = \varepsilon_0(\overleftrightarrow{\chi}_e + 1)$, is a measure of the polarizability of the medium in response to an applied electric field, and $\overleftrightarrow{\varepsilon}$ is the relative permittivity or dielectric function of the material. In a similar manner, we define a linear relation between **H** and **B**,

$$\mathbf{H} = \left[\overleftrightarrow{\mu}\mu_0\right]^{-1} \mathbf{B} - \mathbf{M} \,, \tag{2.7}$$

with the permeability $\overleftrightarrow{\mu}\mu_0$ describing the material's ability to permit a magnetic flux within it, where $\overleftrightarrow{\mu}$ is the relative permeability of the material. Throughout this thesis we consider materials that respond weakly to external magnetic fields within the operating frequency range, which allows us to set $\mathbf{M} = 0$. Equations (2.6) and (2.7) comprise the constitutive relations of a system devoid of any magnetoelectric coupling. If we now allow for the applied electric field to cause an in-phase magnetic field and vice versa, then the constitutive relations that give the ith component of **D** and **B** in the space-time domain take the form

$$D_i(\mathbf{r}, t) = \sum_j \int d^3r' \int dt' \left[\varepsilon_0 \varepsilon_{ij}(\mathbf{r}', t', \mathbf{r}, t) E_j(\mathbf{r}', t') + \xi_{ij}(\mathbf{r}', t', \mathbf{r}, t) H_j(\mathbf{r}', t')\right] \,, \tag{2.8a}$$

$$B_i(\mathbf{r}, t) = \sum_j \int d^3r' \int dt' \left[\zeta_{ij}(\mathbf{r}', t', \mathbf{r}, t) E_j(\mathbf{r}', t') + \mu_0 \mu_{ij}(\mathbf{r}', t', \mathbf{r}, t) H_j(\mathbf{r}', t')\right] \,, \tag{2.8b}$$

[3] Equations (2.5)–(2.7) are written in the frequency domain.

where $\overleftrightarrow{\xi}$ and $\overleftrightarrow{\zeta}$ are the magnetoelectric coupling tensors [2]. The above lengthy relations express that, in the most general case, the system's response measured at a point $(\mathbf{r}, t)$ in space-time depends on the applied field at all other points $(\mathbf{r}', t')$. Depending on the size and the material properties of the systems involved, Eqs. (2.8) can be simplified significantly, as we discuss in what follows.

A first step towards the simplification of the constitutive relations is to consider time homogeneity, which modifies the response tensors as

$$\overleftrightarrow{\varepsilon}(\mathbf{r}', t', \mathbf{r}, t) = \overleftrightarrow{\varepsilon}(\mathbf{r}', \mathbf{r}, t - t')\,, \tag{2.9a}$$

$$\overleftrightarrow{\mu}(\mathbf{r}', t', \mathbf{r}, t) = \overleftrightarrow{\mu}(\mathbf{r}', \mathbf{r}, t - t')\,, \tag{2.9b}$$

$$\overleftrightarrow{\xi}(\mathbf{r}', t', \mathbf{r}, t) = \overleftrightarrow{\xi}(\mathbf{r}', \mathbf{r}, t - t')\,, \tag{2.9c}$$

$$\overleftrightarrow{\zeta}(\mathbf{r}', t', \mathbf{r}, t) = \overleftrightarrow{\zeta}(\mathbf{r}', \mathbf{r}, t - t')\,. \tag{2.9d}$$

Moreover, *causality* dictates that the response of a system at time t depends on the value of the applied field only prior to t, and as a consequence all response functions are zero at times $t < t'$. Furthermore, a large part of this thesis is performed within LRA, meaning that the response field at point $\mathbf{r}$ depends only on the generating field at the same point. It is safe to assume local response, as long as the characteristic size of the structure under study is small enough compared to intrinsic length scales. We address the implications of *nonlocality* in Sect. 2.3.5. In LRA, the response tensors take the form

$$\overleftrightarrow{\varepsilon}(\mathbf{r}', t', \mathbf{r}, t) = \overleftrightarrow{\varepsilon}(\mathbf{r}, t - t')\delta(\mathbf{r} - \mathbf{r}')\,, \tag{2.10a}$$

$$\overleftrightarrow{\mu}(\mathbf{r}', t', \mathbf{r}, t) = \overleftrightarrow{\mu}(\mathbf{r}, t - t')\delta(\mathbf{r} - \mathbf{r}')\,, \tag{2.10b}$$

$$\overleftrightarrow{\xi}(\mathbf{r}', t', \mathbf{r}, t) = \overleftrightarrow{\xi}(\mathbf{r}, t - t')\delta(\mathbf{r} - \mathbf{r}')\,, \tag{2.10c}$$

$$\overleftrightarrow{\zeta}(\mathbf{r}', t', \mathbf{r}, t) = \overleftrightarrow{\zeta}(\mathbf{r}, t - t')\delta(\mathbf{r} - \mathbf{r}')\,. \tag{2.10d}$$

These assumptions already simplify significantly the constitutive relations (2.8) to

$$D_i(\mathbf{r}, t) = \sum_j \int dt' \left[\varepsilon_0 \varepsilon_{ij}(\mathbf{r}, t - t') E_j(\mathbf{r}, t') + \xi_{ij}(\mathbf{r}, t - t') H_j(\mathbf{r}, t')\right]\,, \tag{2.11a}$$

$$B_i(\mathbf{r}, t) = \sum_j \int dt' \left[\zeta_{ij}(\mathbf{r}, t - t') E_j(\mathbf{r}, t') + \mu_0 \mu_{ij}(\mathbf{r}, t - t') H_j(\mathbf{r}, t')\right]\,. \tag{2.11b}$$

By applying a Fourier transform, Eqs. (2.11) can be written in the frequency domain as

$$D_i(\mathbf{r}, \omega) = \sum_j \left[\varepsilon_0 \varepsilon_{ij}(\mathbf{r}, \omega) E_j(\mathbf{r}, \omega) + \xi_{ij}(\mathbf{r}, \omega) H_j(\mathbf{r}, \omega)\right] , \tag{2.12a}$$

$$B_i(\mathbf{r}, \omega) = \sum_j \left[\zeta_{ij}(\mathbf{r}, \omega) E_j(\mathbf{r}, \omega) + \mu_0 \mu_{ij}(\mathbf{r}, \omega) H_j(\mathbf{r}, \omega)\right] , \tag{2.12b}$$

where ω is the angular frequency, and

$$F(\mathbf{r}, \omega) = \int_{-\infty}^{+\infty} dt\, \mathrm{e}^{i\omega t} F(\mathbf{r}, t) , \qquad F = E_i, D_i, \varepsilon_{ij}, \mu_{ij}, \xi_{ij}, \zeta_{ij} . \tag{2.13}$$

Until this point, the constitutive relations are quite general (we have only assumed a local and linear response) and describe any *bi-anisotropic* medium. As we discuss in Sect. 2.2, further assumptions regarding the response of the medium lead to different families of materials.

2.1.3 Boundary Conditions at an Interface

When solving electrodynamic problems one is very often faced with the task of determining the EM fields in different media with response functions that terminate abruptly, thus introducing a discontinuity at the boundary separating, say, medium 1 from medium 2. The fields within the two domains are generally different, but their components exactly at the boundary must be related, for Maxwell's equations to hold in all space. Despite their elegance, Maxwell's equations in the differential form of Eqs. (2.2) are local and inconvenient for determining the (dis)continuity of the fields across the boundary. This can be remedied by bringing the expressions into an integral form. By imposing the divergence (Gauss) theorem[4] to Eqs. (2.2a) and (2.2d) and the curl (Stokes) theorem[5] to Eqs. (2.2b) and (2.2c), we obtain

Gauss' law
$$\oint_S da\, \hat{\mathbf{n}} \cdot \mathbf{D} = \int_V d^3 r\, \rho_{\mathrm{f}} , \tag{2.14a}$$

Ampère-Maxwell's law
$$\oint_C d\mathbf{l} \cdot \mathbf{H} = \int_{S'} da\, \hat{\mathbf{n}}' \cdot \left[\mathbf{J}_{\mathrm{f}} + \frac{\partial \mathbf{D}}{\partial t}\right] , \tag{2.14b}$$

Faraday's law
$$\oint_C d\mathbf{l} \cdot \mathbf{E} = -\int_{S'} da\, \hat{\mathbf{n}}' \cdot \frac{\partial \mathbf{B}}{\partial t} , \tag{2.14c}$$

Absence of magnetic monopoles
$$\oint_S da\, \hat{\mathbf{n}} \cdot \mathbf{B} = 0 . \tag{2.14d}$$

[4] The divergence theorem relates the flux of a vector field $\mathbf{F}$ through a closed surface S to the divergence of the field in the volume V enclosed. In math notation it states that $\int_V d^3 r\, (\nabla \cdot \mathbf{F}) = \oint_S da\, \hat{\mathbf{n}} \cdot \mathbf{F}$.

[5] The curl theorem relates the integral of the curl of the vector field $\mathbf{F}$ over some surface S', to the line integral of the vector field around the boundary of the surface. In math notation it states that $\int_{S'} da\, \hat{\mathbf{n}} \cdot (\nabla \times \mathbf{F}) = \oint_C d\mathbf{l} \cdot \mathbf{F}$.

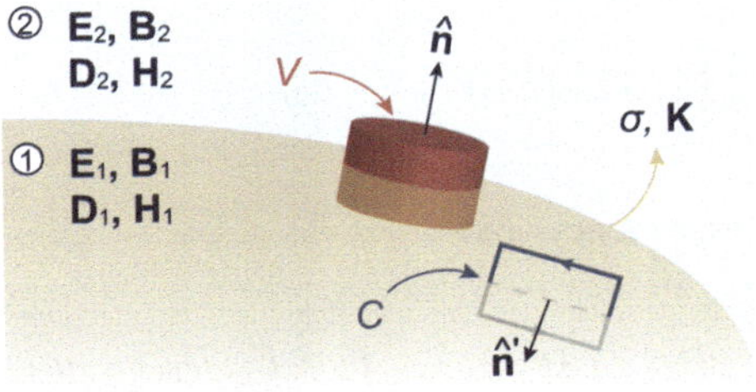

Fig. 2.1 Boundary conditions at an interface. Media 1 and 2 are separated by a boundary that carries surface charge density σ, and surface current density $\mathbf{K}$. The relations linking the EM fields in the two domains are determined by considering a small Gaussian pillbox of volume V (red cylinder) with its top plane oriented tangentially to the boundary, and a Stokesian loop (contour C) whose plane is oriented normally

Here, V denotes a finite volume in space, C a closed contour, S and S' a closed and an open surface respectively, da is the infinitesimal element of area, and $d\mathbf{l}$ is the infinitesimal line element on the contour. Furthermore, $\hat{\mathbf{n}}$ denotes a unit vector normal to da pointing outward from the enclosed volume, and $\hat{\mathbf{n}}'$ a unit vector normal to da pointing in the direction given by the right-hand rule from the sense of integration around contour C, as depicted in Fig. 2.1.

By using the Gaussian pillbox argument on the boundary [1], and Eqs. (2.14a) and (2.14d), we determine the relation linking the normal components of $\mathbf{D}$ and $\mathbf{B}$ on either side of the interface. Similarly, by using a Stokesian loop perpendicular to the boundary, and Eqs. (2.14b) and (2.14c), we arrive at the relation linking the tangential components of $\mathbf{E}$ and $\mathbf{H}$,

Box 2.2 | Boundary conditions at an interface

Normal field components

$$\hat{\mathbf{n}} \cdot (\mathbf{D}_2 - \mathbf{D}_1) = \sigma \,, \tag{2.15a}$$

$$\hat{\mathbf{n}} \cdot (\mathbf{B}_2 - \mathbf{B}_1) = 0 \,, \tag{2.15b}$$

Tangential field components

$$\hat{\mathbf{n}} \times (\mathbf{E}_2 - \mathbf{E}_1) = 0 \,, \tag{2.15c}$$

$$\hat{\mathbf{n}} \times (\mathbf{H}_2 - \mathbf{H}_1) = \mathbf{K} \,, \tag{2.15d}$$

where σ and $\mathbf{K}$ are the surface charge and current density, respectively. In this thesis, we are not involved with systems sustaining surface current densities ($\mathbf{K} = 0$), and we therefore utilize the continuity of the tangential components of both $\mathbf{E}$ and $\mathbf{H}$ as self-evident.

2.1.4 *Wave Equation of Time-Harmonic Fields*

Considering linear response, Maxwell's equations form a set of linear equations. The *superposition principle* allows us to express any solution to the system as a linear combination of monochromatic waves. For monochromatic time-harmonic fields of the form

$$\mathbf{E}(\mathbf{r}, t) = \mathrm{Re}\{\mathbf{E}(\mathbf{r}, \omega)\mathrm{e}^{-i\omega t}\} , \tag{2.16a}$$

$$\mathbf{H}(\mathbf{r}, t) = \mathrm{Re}\{\mathbf{H}(\mathbf{r}, \omega)\mathrm{e}^{-i\omega t}\} , \tag{2.16b}$$

the macroscopic Maxwell equations in the frequency domain read

$$\nabla \cdot \mathbf{D} = \rho_\mathrm{f} , \tag{2.17a}$$

$$\nabla \times \mathbf{H} = -i\omega \mathbf{D} + \mathbf{J}_\mathrm{f} , \tag{2.17b}$$

$$\nabla \times \mathbf{E} = i\omega \mathbf{B} , \tag{2.17c}$$

$$\nabla \cdot \mathbf{B} = 0 . \tag{2.17d}$$

Let us assume for now the constitutive relations of a simple isotropic medium (we discuss thoroughly different classes of materials in Sect. 2.2),

$$\mathbf{D} = \varepsilon\varepsilon_0 \mathbf{E} , \quad \mathbf{B} = \mu\mu_0 \mathbf{H} . \tag{2.18}$$

Acting with the curl operator on Faraday's law (2.17c), while also using Eqs. (2.17b) and (2.16), we find

$$-\nabla \times \nabla \times \mathbf{E}(\mathbf{r}) + \frac{\omega^2}{c^2}\varepsilon\mu \mathbf{E}(\mathbf{r}) = -i\omega\mu\mu_0 \mathbf{J}(\mathbf{r}) , \tag{2.19}$$

where $c = 1/\sqrt{(\varepsilon_0\mu_0)}$ is the speed of light in vacuum. The form of Eq. (2.19) suggests to relate the electric field and the current source via the dyadic Green's function $\overleftrightarrow{\mathscr{G}}$.

Box 2.3 | Dyadic Green's function in homogeneous space

In homogeneous space the dyadic Green's function in matrix notation reads

$$\overleftrightarrow{\mathscr{G}}(\mathbf{r}, \mathbf{r}') = \left(\mathbf{I} + \frac{\nabla\nabla}{k^2}\right) \frac{\mathrm{e}^{ik|\mathbf{r}-\mathbf{r}'|}}{4\pi|\mathbf{r}-\mathbf{r}'|} , \tag{2.20}$$

and fulfills the equation

$$-\nabla \times \nabla \times \overleftrightarrow{\mathscr{G}}(\mathbf{r}, \mathbf{r}') + k^2 \overleftrightarrow{\mathscr{G}}(\mathbf{r}, \mathbf{r}') = -\delta(\mathbf{r}-\mathbf{r}')\mathbf{I} , \tag{2.21}$$

where **I** is the unit matrix. Then the electric field can be expressed as

$$\mathbf{E}(\mathbf{r}) = i\omega\mu\mu_0 \int \overleftrightarrow{\mathscr{G}}(\mathbf{r}, \mathbf{r}') \cdot \mathbf{J}(\mathbf{r}') d\mathbf{r}' \,. \tag{2.22}$$

Returning to Eq. (2.19), and using Eq. (2.17a), we can construct the inhomogeneous Helmholtz equation for the electric field

$$\left(\nabla^2 + \frac{\omega^2}{c^2}\varepsilon\mu\right)\mathbf{E} = -i\omega\mu\mu_0\mathbf{J}_\mathrm{f} + \frac{1}{\varepsilon\varepsilon_0}\nabla\rho_\mathrm{f} \,. \tag{2.23}$$

Here ∇^2 denotes the vector Laplacian, $\nabla^2 = \nabla(\nabla \cdot \mathbf{A}) - \nabla \times \nabla \times \mathbf{A}$. In the absence of free charges and currents, we obtain the homogeneous Helmholtz equation

Box 2.4 | Helmholtz equation for the electric field

$$\left(\nabla^2 + \frac{\omega^2}{c^2}\varepsilon\mu\right)\mathbf{E} = 0 \,. \tag{2.24}$$

A similar equation holds for the magnetic field **H**.

2.1.5 *Electromagnetic Potentials*

From Eq. (2.17d), together with the vector identity $\nabla \cdot (\nabla \times \mathbf{A}) = 0$, one can introduce a vector potential **A**, so that

$$\mathbf{B} = \nabla \times \mathbf{A} \,. \tag{2.25}$$

We can then write Eq. (2.17c) as

$$\nabla \times (\mathbf{E} - i\omega\mathbf{A}) = 0 \,. \tag{2.26}$$

Considering that every scalar potential φ obeys the identity $\nabla \times (\nabla\varphi) = 0$, we can express the electric field as

$$\mathbf{E} = i\omega\mathbf{A} - \nabla\varphi \,. \tag{2.27}$$

The EM field can, therefore, be constructed by the vector potential **A** and the scalar potential φ. Clearly, the potentials are also time-harmonic and satisfy, in the general case, an inhomogeneous Helmholtz equation, which can be constructed once we pick the appropriate gauge.[6] For the electrodynamic problems addressed in this thesis, we pick the *Lorenz gauge*

$$\nabla \cdot \mathbf{A} = -\varepsilon\varepsilon_0\mu\mu_0 \frac{\partial \varphi}{\partial t} = i\frac{\omega}{c^2}\varepsilon\mu\varphi \,. \tag{2.28}$$

If we now insert the Lorenz gauge into Eq. (2.27), and the constitutive relations of Eqs. (2.18) into Gauss's (2.17a) and Faraday's (2.17c) laws, we obtain, respectively,

$$\left(\nabla^2 + \frac{\omega^2}{c^2}\varepsilon\mu\right)\varphi = -\frac{\rho_\mathrm{f}}{\varepsilon\varepsilon_0} \,, \tag{2.29a}$$

$$\left(\nabla^2 + \frac{\omega^2}{c^2}\varepsilon\mu\right)\mathbf{A} = -\mu\mu_0\mathbf{J}_\mathrm{f} \,. \tag{2.29b}$$

2.1.6 Plane Waves

A set of solutions to the homogeneous Helmholtz equation (2.24) is given by the time-harmonic plane waves, propagating along the direction dictated by the wavevector $\mathbf{k} = k\hat{\mathbf{k}}$, i.e.

$$\mathbf{E}(\mathbf{r}, t) = \mathbf{E}_0(\mathbf{k})\mathrm{e}^{i\mathbf{k}\cdot\mathbf{r}-i\omega t} \,, \tag{2.30a}$$

$$\mathbf{H}(\mathbf{r}, t) = \mathbf{H}_0(\mathbf{k})\mathrm{e}^{i\mathbf{k}\cdot\mathbf{r}-i\omega t} \,, \tag{2.30b}$$

where $\mathbf{E}_0$ and $\mathbf{H}_0$ are constant vectors that determine the polarization of the fields. After applying a time-space Fourier transform, and once inserting the plane-wave solution (2.30a) in Eq. (2.24), the wave equation can be written as

$$\mathbf{k}[\mathbf{k} \cdot \mathbf{E}_0(\mathbf{k}, \omega)] + \frac{\omega^2}{c^2}\varepsilon\mu\mathbf{E}_0(\mathbf{k}, \omega) = k^2\mathbf{E}_0(\mathbf{k}, \omega) \,. \tag{2.31}$$

This form allows us to divide the solution in two classes, according to the electric field polarization: *transverse*, with polarization perpendicular to the propagation direction ($\mathbf{k} \cdot \mathbf{E}_0 = 0$), which yields

$$\frac{\omega^2}{c^2}\varepsilon\mu\mathbf{E}_0(\mathbf{k}, \omega) = k^2\mathbf{E}_0(\mathbf{k}, \omega) \,, \tag{2.32}$$

and *longitudinal*, with parallel polarization ($\mathbf{k} \cdot \mathbf{E}_0 = kE_0$), giving

[6] There is a level of arbitrariness in the definition of $\mathbf{A}$ and φ. Different choices for the potentials that leave Eqs. (2.25) and (2.27) unchanged correspond to different gauges. The most popular choices are the Coulomb and the Lorenz gauge. The former sets $\nabla \cdot \mathbf{A} = 0$, and is most often used when no sources are present, but generally raises problems of causality [1].

$$\frac{\omega^2}{c^2}\varepsilon\mu\mathbf{E}_0(\mathbf{k},\omega) = 0\ . \tag{2.33}$$

Box 2.5 | Transverse and longitudinal waves

Propagation of transverse waves is only possible when the wavenumber is equivalent to

$$k = \frac{\omega}{c}\sqrt{\varepsilon\mu}\ , \tag{2.34}$$

whereas longitudinal waves arise when

$$\varepsilon = 0\ , \tag{2.35}$$

since materials with zero permeability do not exist.

Let us now focus exclusively on the transverse waves. Faraday's law (2.17c) reveals that $\mathbf{E}_0$ and $\mathbf{H}_0$ are in phase and mutually perpendicular, or

$$\mathbf{H}_0 = \frac{1}{\omega\mu\mu_0}\hat{\mathbf{k}} \times \mathbf{E}_0\ . \tag{2.36}$$

The polarization vectors of plane waves are confined to the plane normal to the propagation direction. If, furthermore, they are always along a specific direction, say $\hat{\mathbf{e}}$ for the electric field, we say that the wave is *linearly polarized*. In the general case, the polarization vector can be decomposed into different components that determine the type of polarization of the EM wave. Let us assume a vector

$$\mathbf{E}_0 = \hat{\mathbf{e}}_1 E_1 + \hat{\mathbf{e}}_2 E_2\ , \tag{2.37}$$

where E_1, E_2 are complex numbers to allow the possibility of phase difference between waves of different linear polarization. If E_1 and E_2 have the same phase, then Eq. (2.37) represents a linearly polarized wave. In any other case, where E_1 and E_2 have a phase difference while $|E_1| \neq |E_2|$, Eq. (2.37) represents an *elliptically polarized* wave. In the special case that E_1 and E_2 have the same magnitude $|E_1| = |E_2| = E_0$, but differ by a phase difference of $\pi/2$, i.e. $\mathbf{E}_0 = E_0(\hat{\mathbf{e}}_1 \pm i\hat{\mathbf{e}}_2)$, the wave is *circularly polarized*. As the name suggests, the above equation represents a plane wave whose polarization vector rotates on the $\hat{\mathbf{e}}_1\hat{\mathbf{e}}_2$ plane with a constant magnitude. The (+) combination corresponds to a clockwise rotating wave (with respect to an observer that is facing the wave as it approaches), called *left circularly polarized* (LCP) wave, while (−) corresponds to a counterclockwise rotating wave, called *right circularly polarized* (RCP) wave.

The dielectric function and, consequently, the wavenumber can take complex values $k = k' + ik''$. The real part of the wavenumber is associated with the propagation speed of the phase fronts of the wave (*phase velocity*)

$$v_{\mathrm{ph}} = \frac{\omega}{k'}, \tag{2.38}$$

while the imaginary part describes attenuation of the wave amplitude as it propagates through space. In other words, the imaginary part of the dielectric function embodies the loss mechanisms of the material that remove energy from an EM wave propagating therein.

2.2 Material Properties

In Sect. 2.1.2 we mentioned that the optical properties of a material are encoded within its permittivity and permeability. Here we touch upon a few families of materials that are relevant to this thesis; we present the form of the corresponding optical parameters ε and μ, and how the general constitutive relations (2.8) simplify in each case. It is useful, at this point, to introduce the concept of isotropy, referring to the property of a medium whose EM properties are the same in all directions.

Simple Isotropic Media The magnetoelectric coupling constants are zero ($\xi, \zeta = 0$), and the optical parameters acquire scalar values $\overleftrightarrow{\varepsilon} = \varepsilon$ and $\overleftrightarrow{\mu} = \mu$. The constitutive relations take their simplest form

$$\mathbf{D} = \varepsilon\varepsilon_0\mathbf{E}, \quad \mathbf{B} = \mu\mu_0\mathbf{H}. \tag{2.39}$$

In a simple isotropic medium, an applied electric field induces charge separation therein, creating an electric dipole moment. In analogous manner, in magnetic materials ($\mu \neq 1$) an applied magnetic field creates a magnetic dipole moment.

Bi-isotropic Media All response functions are isotropic (scalar) and the coupling constants are related to the Tellegen, χ, and chirality, or Pasteur, κ parameter [2]

$$\xi = \frac{\chi + i\kappa}{c}, \quad \zeta = \frac{\chi - i\kappa}{c}. \tag{2.40}$$

If, additionally, the Tellegen parameter is zero ($\chi = 0$ and $\kappa \neq 0$), we speak of a *chiral* or a *Pasteur* medium, whereas in the opposite case ($\kappa = 0$ and $\chi \neq 0$) we speak of a *Tellegen* medium. Inside a chiral medium the constitutive relations take the form

$$\mathbf{D} = \varepsilon\varepsilon_0\mathbf{E} + \frac{i\kappa}{c}\mathbf{H}, \quad \mathbf{B} = -\frac{i\kappa}{c}\mathbf{E} + \mu\mu_0\mathbf{H}. \tag{2.41}$$

In this thesis we are only concerned with chiral bi-isotropic media, for which the response of the **D**-field to the **H**-field is delayed by a phase $\exp(i\pi/2)$. The concept of magnetoelectric coupling can be understood intuitively if we think of the elements comprising the chiral medium as helices. An applied electric field polarizes the medium, creating an electric dipole moment, but, at the same time, the shape of the helix forces the charges to move circularly in addition to the linear motion. This electric current loop induces a magnetic dipole moment, introduced by the Pasteur parameter κ in the constitutive relations (2.41). Both polarizations are induced also when an external magnetic field is applied.

Gyrotropic Media Gyrotropic media are a subgroup of anisotropic media, that exhibit anisotropy when subjected to a static magnetic field. Assuming a negligible magnetoelectric coupling ($\xi, \zeta = 0$), they can be further classified into *gyroelectric* and *gyromagnetic* media. Gyroelectric properties are found in some rare-earth garnets, such as yttrium iron garnet, and they are described by a scalar permeability and tensorial permittivity of the form [2]

$$\overleftrightarrow{\varepsilon} = \varepsilon_z \begin{pmatrix} \varepsilon_x & -i\varepsilon_{xy} & 0 \\ i\varepsilon_{xy} & \varepsilon_x & 0 \\ 0 & 0 & 1 \end{pmatrix}, \tag{2.42}$$

for a magnetic field oriented along the z axis. Similarly, gyromagnetic media are described by a scalar permittivity and tensorial permeability

$$\overleftrightarrow{\mu} = \mu_z \begin{pmatrix} \mu_x & -i\mu_{xy} & 0 \\ i\mu_{xy} & \mu_x & 0 \\ 0 & 0 & 1 \end{pmatrix}. \tag{2.43}$$

Of course, certain materials can combine more than one of the properties mentioned above, as indicated by the general expressions of Eqs. (2.12).

In the present thesis we will be working with nonmagnetic metals and high-index dielectrics in the optical regime, where the relative permeability can be taken as constant and equal to unity. In most expressions we explicitly show the dependence on μ for reasons of generality, but all results have been derived assuming $\mu = 1$.

2.2.1 *Optical Activity and Chirality*

Optical activity pertains to phenomena that can be attributed to distinct responses of a system to LCP and RCP light [3]. Bi-isotropic and gyrotropic media, as defined in the previous section, exhibit this interesting behavior. Optical activity manifests, in particular, through two phenomena, referred to as *optical rotation* (OR) and *circular dichroism* (CD). OR is the property of optically active media to rotate the plane of polarization of an incident plane harmonic wave, unless it is circularly polarized. CD refers to the different absorption suffered by RCP and LCP waves traveling through the medium. Various types of optical activity exist, which naturally reflect on the exact form of the response functions. Gyrotropic media, for instance, exhibit magnetically induced optical activity (magnetochirality), whereas chiral media feature natural optical activity.

Natural optical activity was first observed as early as in the beginning of the 19th century (Arago, 1811) in quartz crystals, and later in several organic liquids, such as turpentine and aqueous solutions of sugar. It was quickly realized that the peculiar optical response of these systems stems from their chiral molecular or crystal structure. *Chirality* refers to the characteristic of their elementary building blocks that they are not superimposable onto their mirror image under any rotation or translation (see Fig. 2.2a). The two mirror versions of a chiral object are called *enantiomers*. Apart from the *constitutional* chirality, meaning that the chirality of the object is traced back to its chiral molecules, non-chiral systems can be arranged in a geometry that does not obey mirror symmetry, thereby exhibiting what is termed *configurational* chirality [4]. This can be accomplished, for example, by arranging metallic NPs in a helical configuration, as illustrated in Fig. 2.2b.

Being a common characteristic of biomolecules and massively determining their chemical functions, chirality has remained the subject of intense study for several decades. Chiral matter can be best detected and analyzed through its interaction with other chiral objects. In the field of photonics, circularly polarized light comprises the most fundamental chiral entity, and thus the most common probe for the characterization of enantiomers. It is clear that chirality in matter is inextricably connected to optical activity, and therefore typically quantified via CD measurements.

Another relevant quantity often used to characterize the local density of optical chirality (handedness) of an EM field, and potentially enhance it, is the optical chirality density C. It was originally introduced by Lipkin [5], defined as

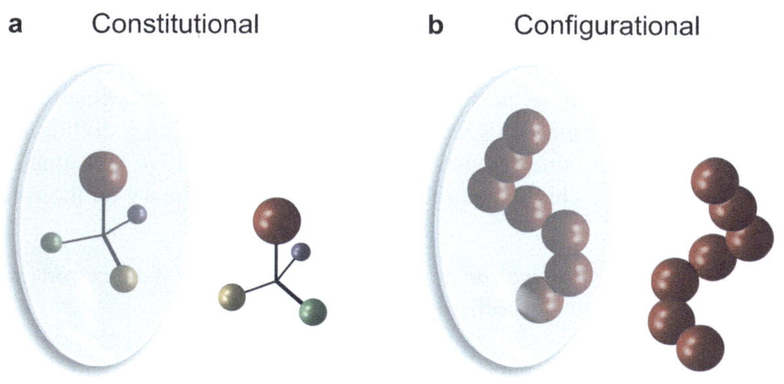

Fig. 2.2 Types of chirality in matter. **a** Constitutional chirality due to chiral elementary building blocks in a non-chiral arrangement. **b** Configurational chirality due to non-chiral elementary building blocks in a chiral arrangement

$$C(\mathbf{r},\omega) = \frac{1}{2}\left[\mathbf{E}\cdot(\nabla\times\mathbf{D}) + \mathbf{H}\cdot(\nabla\times\mathbf{B})\right] , \tag{2.44}$$

who conceived it as a mathematical entity without any physical significance. It was only recently that Tang and Cohen ascribed an actual physical interpretation to it, describing it as a measure of "screwiness" of the electric and magnetic field lines [6]. In a lossless and nondispersive medium (real and constant ε and μ) the time-averaged optical chirality density[7] takes the form

$$C(\mathbf{r},\omega) = \frac{\omega\varepsilon\mu}{2c^2}\mathrm{Im}\left\{\mathbf{E}(\mathbf{r},\omega)\cdot\mathbf{H}^*(\mathbf{r},\omega)\right\} . \tag{2.45}$$

Inside a chiral medium, the chirality density is related to the local absorption $A^{\pm}$ of L/RCP waves via the expression [7]

$$A^{\pm} = \frac{\omega}{2c^2}\left[\mathrm{Im}\{\varepsilon\}\,|\mathbf{E}|^2 + \mathrm{Im}\{\mu\}\,|\mathbf{H}|^2\right] \pm \frac{\omega}{c}\mathrm{Im}\{\kappa\}\mathrm{Im}\left\{\mathbf{E}\cdot\mathbf{H}^*\right\} . \tag{2.46}$$

Equation (2.46) reveals that, in addition to achiral absorption (first term), chiral absorption (second term) also contributes to the overall absorption of the EM field. According to Eq. (2.46), the CD due to chiral absorption is expected to be proportional to the chirality density, as

$$\mathrm{CD} \propto \mathrm{Im}\{\kappa\}\, C . \tag{2.47}$$

2.2.2 *The Dielectric Function*

2.2.2.1 Electrons in Metals and the Drude Model

The defining characteristic that distinguishes metals from other materials is their ability to conduct electricity, owing to their mobile electrons. Shortly after the discovery of the electron by J. J. Thomson in 1896, Paul Drude realized that he could treat those mobile electrons as an electron gas and apply Boltzman's kinetic theory of gases.[8] He assumed that electrons move in straight lines, until they collide with one another or with a background consisting of much heavier positive ions, that must be there to compensate for charge neutrality. Drude based his theory on the following assumptions [8]:

1. The *free electron approximation* neglects the interaction of the electrons with the ionic background between collisions.

---

[7] From now on, we use the shorter term chirality density to refer to the time-averaged optical chirality density.

[8] In the kinetic theory of gases the molecules of a gas are considered identical particles that move in straight trajectories until they collide with one another, while intermolecular forces between collisions are neglected.

2. The *independent electron approximation* neglects electron–electron interaction between collisions.
3. The electron–ion collision is the important scattering mechanism and it occurs instantaneously.
4. The scattering time between collisions with the other electrons is defined by a parameter τ. The probability of scattering within a time interval dt is dt/τ.
5. Electrons achieve thermal equilibrium with their environment only through collisions.

Consider a single conduction electron of charge $-e$ and mass m_e, moving with velocity $\mathbf{v}$ inside a time-harmonic electric field $\mathbf{E}$. Since it is a charged particle, it experiences a force $\mathbf{F} = -e\mathbf{E}$ [Eq. (2.4)]. According to the assumptions above, its momentum $\mathbf{p}$ is determined by the equation of motion

$$\frac{d\mathbf{p}}{dt} = -e\mathbf{E} - \frac{\mathbf{p}}{\tau} \,. \tag{2.48}$$

The Fourier transform of the equation of motion yields

$$\mathbf{p}(\omega) = \frac{-e}{1/\tau - i\omega}\mathbf{E}(\omega) \,. \tag{2.49}$$

Given an electron density n, the current density is $\mathbf{J} = -en\mathbf{v}$, leading to

$$\mathbf{J}(\omega) = -\frac{en}{m_e}\mathbf{p}(\omega) = \frac{e^2 n}{m_e}\frac{\mathbf{E}(\omega)}{1/\tau - i\omega} \,. \tag{2.50}$$

Using Ohm's law, $\mathbf{J}(\omega) = \sigma(\omega)\mathbf{E}(\omega)$, we obtain the Drude conductivity of the metal

$$\sigma(\omega) = \frac{\sigma_0}{1 - i\omega\tau} \,, \tag{2.51}$$

with $\sigma_0 = e^2\tau n/m_e$.

By introducing the current density into the inhomogeneous wave equation (2.23), while also considering that we have explicitly included the losses due to conduction by inserting Ohm's law into the wave equation (so here $\mu = 1$ and $\varepsilon = 1$), we obtain

$$k^2 = \frac{\omega^2}{c^2} + i\mu_0\sigma\omega \,. \tag{2.52}$$

If, on the other hand, we do not insert Ohm's law explicitly, but attribute instead all the properties of the medium to the permittivity, we can identify the quantity

$$\varepsilon(\omega) = 1 + \frac{i\sigma}{\varepsilon_0\omega} \tag{2.53}$$

as a new dielectric function. Together with Eq. (2.51), we arrive at the Drude model for the dielectric function of metals

$$\varepsilon(\omega) = 1 - \frac{\omega_{\mathrm{p}}^2}{\omega(\omega + i\gamma)} \,, \tag{2.54}$$

where $\gamma = 1/\tau$ it the damping rate and $\omega_{\mathrm{p}} = \sqrt{e^2 n / m_e \varepsilon_0}$ is the plasma frequency. Below the plasma frequency the real part of the dielectric function is negative, which is associated with metallic behavior.

Despite its simplicity, Drude's model was successful in describing several transport properties of electrons in metals (Hall effect, Wiedemann-Franz law), but failed in predicting others with good agreement (Peltier and Seeback coefficients). These shortcomings resulted from the fact that Drude's theory was based on the classical Boltzman statistics for the electron gas. After the postulation of the exclusion principle[9] in 1925 by Pauli and the development of Fermi-Dirac statistics in 1926 [9], Sommerfeld incorporated the latter, introducing the idea that electrons occupy discrete energy levels forming their *band structure* [10]. Overall, the electric and optical properties of metals are determined by their band structure, given by the *dispersion relation* between the electron energies $E(k)$ and the wavenumber k. For simple metals, it can be approximated by the parabolic dispersion relation of free electrons [8]

$$E(k) = \frac{\hbar^2 k^2}{2 m_e} \,, \tag{2.55}$$

where $\hbar$ is the reduced Planck constant. All states are filled up to the *Fermi energy* level E_{F}, while the energy levels below E_{F} are occupied by electrons that are more tightly bound to the ions. The Drude model can be modified to account for contributions of the bound electrons by replacing the unity in Eq. (2.54) by a background relative permittivity $\varepsilon_\infty(\omega)$, as

Box 2.6 | Relative permittivity in the Drude model

$$\varepsilon(\omega) = \varepsilon_\infty(\omega) - \frac{\omega_{\mathrm{p}}^2}{\omega(\omega + i\gamma)} \,. \tag{2.56}$$

Noble metals, like silver and gold, in addition to free electrons, host strongly localized electrons in d-band states (also f-band in gold) below the Fermi energy. When excited by a photon of energy larger than their energy distance, the d-band electrons can partake in *interband transitions*, namely they can be promoted to energy states above the Fermi level. When the excitation energy does not suffice for an interband

[9] The Pauli exclusion principle states that two or more fermions cannot simultaneously occupy the same quantum state within a quantum system.

transition, the electrons in the d-band states polarize, leading to *d-band screening*. Both processes translate to large values of ε_∞.

2.2.2.2 Electrons in Dielectrics and the Lorentz Model

The assumption of free electrons in the Drude model is a good enough approximation to describe basic transport properties in metals, but not in nonconducting materials, where strong attractive forces from the ions do not allow them to move freely through the ionic lattice. Once subjected to an electric field, rather than move freely, the electrons are displaced away from the ions, giving rise to an effective dipole. Such a polarizable material, that we will call *dielectric* hereinafter, can be more accurately modelled by the Lorentz model as a collection of harmonic oscillators of eigenfrequencies ω_0. The equation of motion of a bound electron due to a harmonic force that varies in time as $\mathrm{e}^{-i\omega t}$ is given by

$$m_e \frac{d^2\mathbf{x}}{dt^2} = -m_e\gamma \frac{d\mathbf{x}}{dt} - m_e\omega_0^2\mathbf{x} - e\mathbf{E}\,. \tag{2.57}$$

The system will oscillate with the driving frequency ω, resulting in a displacement

$$\mathbf{x} = \frac{e/m_e}{\omega^2 - \omega_0^2 + i\omega\gamma}\mathbf{E}\,, \tag{2.58}$$

and a polarization density of n molecules/oscillators per unit volume

$$\mathbf{P} = -ne\mathbf{x} = -\frac{ne^2/m_e}{\omega^2 - \omega_0^2 + i\omega\gamma}\mathbf{E}\,. \tag{2.59}$$

From Eq. (2.5) we can derive the Lorentz dielectric function

$$\varepsilon(\omega) = 1 - \frac{ne^2}{m_e\varepsilon_0}\frac{1}{\omega^2 - \omega_0^2 + i\omega\gamma}\,. \tag{2.60}$$

Evidently, the Lorentz model reproduces the Drude dielectric function, once we remove the restoring force ($\omega_0 \to 0$). We may substitute the prefactor $ne^2/(m_e\varepsilon_0)$ for an unknown oscillator strength f times ω_0^2 to facilitate fitting to experimental data, as well as replace once again the unity with a background relative permittivity $\varepsilon_\infty(\omega)$, as

Box 2.7 | Relative permittivity in the Lorentz model

$$\varepsilon(\omega) = \varepsilon_\infty(\omega) - \frac{f\omega_0^2}{\omega^2 - \omega_0^2 + i\omega\gamma}, \tag{2.61}$$

to account for other contributions of bound electrons.

2.2.2.3 Electrons in Nonlocal Metals and the Hydrodynamic Drude Model

The Drude model of Eq. (2.56) describes the local response of the metal. As we discuss in detail in Sect. 2.3.5, one way to incorporate nonlocal effects within LRA is via the hydrodynamic Drude model (HDM) [11], that treats the electrons of the metal as a compressible electron gas in a fluid hydrodynamic formulation. The basic assumption of HDM is that the electron dynamics can be expressed in terms of the electron density $n(\mathbf{r}, t)$ and the hydrodynamic velocity $\mathbf{v}(\mathbf{r}, t)$. Under the influence of the EM field, the dynamics of the two variables can be described by the continuity equation (2.3), together with the hydrodynamic equation of motion[10] [12]

$$\frac{\partial n}{\partial t} + \nabla \cdot (n\mathbf{v}) = 0\,, \tag{2.62a}$$

$$\frac{\partial \mathbf{v}}{\partial t} + (\mathbf{v}\cdot\nabla)\,\mathbf{v} = -\frac{e}{m_e}(\mathbf{E} + \mathbf{v}\times\mathbf{B}) - \gamma\mathbf{v} - \frac{1}{m_e}\nabla\frac{\delta G[n]}{\delta n}, \tag{2.62b}$$

where $G[n]$ is a total energy functional of the density that contains all interactions and correlations between electrons, and γ is the damping rate. Taking only the kinetic energy of the electrons into account, G simplifies to the Thomas-Fermi functional $T_{\mathrm{TF}}[n]$ [13],

$$G[n(\mathbf{r}, t)] \approx T_{\mathrm{TF}}[n(\mathbf{r}, t)] = \frac{3\hbar^2}{10 m_e}\left(3\pi^2\right)^{2/3}\int d\mathbf{r}\; n(\mathbf{r}, t)^{5/3}\,. \tag{2.63}$$

In this case, the term $\nabla(\delta G[n]/\delta n)$ in Eq. (2.62b) represents a hydrodynamic pressure $\wp$ gradient describing the convective flow of the electron gas. In the absence of an external field, the electron gas is in equilibrium, maintaining a constant density $n = n_0$. For small deviations from equilibrium, the density acquires an additional out-of-equilibrium part n_1, which translates to $n = n_0 + n_1$ and $\wp = \wp_0 + \wp_1$, accordingly. Then the linearized equations of motion take the form

$$\frac{\partial n_1}{\partial t} + n_0\nabla\cdot\mathbf{v} = 0\,, \tag{2.64a}$$

---

[10] Equations (2.62) express essentially the conservation of charge and momentum in the electron gas.

$$\frac{\partial \mathbf{v}}{\partial t}+\frac{e}{m_e}\mathbf{E}+\frac{\beta^2}{n_0}\nabla n_1=-\gamma\mathbf{v}\,, \tag{2.64b}$$

where we have set the hydrodynamic parameter

$$\beta^2=\frac{1}{m_e}\frac{\partial p_1}{\partial n_1}\,. \tag{2.65}$$

Replacing the velocity with the displacement $\mathbf{v}=\partial\mathbf{x}/\partial t$, and considering the time-harmonic plane-wave excitation of Eq. (2.30a), we find

$$-\omega^2\mathbf{x}=\frac{e}{m_e}\mathbf{E}_0-i\gamma\omega\mathbf{x}+\beta^2\mathbf{k}(\mathbf{k}\cdot\mathbf{x})\,. \tag{2.66}$$

Analyzing $\mathbf{x}$ and $\mathbf{E}_0$ into components parallel and perpendicular to $\mathbf{k}$, we obtain the corresponding displacement components

$$x_\perp=\frac{e}{m_e}\frac{1}{\omega(\omega+i\gamma)}E_{0,\perp} \tag{2.67a}$$

$$x_\parallel=\frac{e}{m_e}\frac{1}{\omega(\omega+i\gamma)-\beta^2k_\parallel^2}E_{0,\parallel}\,, \tag{2.67b}$$

where $k_\parallel$ is the longitudinal component of the wavevector. Utilizing again the polarization density $\mathbf{P}$, as in Eq. (2.59), we end up with the dielectric functions connecting the two components of the electric field with the corresponding polarization

Box 2.8 | Relative permittivity components in HDM

$$\varepsilon_\perp(\omega)=\varepsilon_\infty(\omega)-\frac{\omega_\mathrm{p}^2}{\omega(\omega+i\gamma)}\,, \tag{2.68a}$$

$$\varepsilon_\parallel(\omega)=\varepsilon_\infty(\omega)-\frac{\omega_\mathrm{p}^2}{\omega(\omega+i\gamma)-\beta^2k_\parallel^2}\,, \tag{2.68b}$$

In Eqs. (2.68) we have introduced once more the bound-electron contribution ε_∞ by hand. The transversal component of the relative permittivity reproduces the classical Drude model. Therefore, nonlocal corrections enter our theory only through the presence of the longitudinal part of the electric field in the medium.

2.2.2.4 Frequency Dispersion and Loss

Typically, the velocity of propagation of a wave inside a medium—other than vacuum—varies with the frequency of the wave. This property reflects the frequency dependence of the dielectric function $\varepsilon(\omega)$. In this section we will show that frequency dispersion is tightly linked with the losses of the material, described by the imaginary part of the dielectric function, through the so-called *Kramers–Kronig relations*. Here, we briefly show the basic steps of the proof and refer the reader to Refs. [1, 14] for a more detailed derivation.

We showed in Eqs. (2.12) that frequency *dispersion* results from the temporal nonlocality connecting **D** and **E**. Recalling the argument of causality between excitation and response allows us to write the **D** field of Eq. (2.11) in an isotropic medium as

$$\mathbf{D}(\mathbf{r},t) = \varepsilon_0 \mathbf{E}(\mathbf{r},t) + \varepsilon_0 \int_0^\infty d\tau\, \chi_e(\tau)\mathbf{E}(\mathbf{r},t-\tau)\,, \tag{2.69}$$

where $\chi_e(\tau)$ is the Fourier transform of the electric susceptibility. The dielectric function can then be written as

$$\varepsilon(\omega) = 1 + \int_0^\infty d\tau\, \chi_e(\tau) e^{i\omega\tau}\,. \tag{2.70}$$

As a consequence of Eq. (2.70), the real and imaginary parts of the permittivity are interconnected. Taking the complex conjugate of Eq. (2.70), and considering that **D**, **E**, and therefore $\chi_e(\tau)$ take real values, we obtain for complex ω

$$\varepsilon^*(\omega^*) = \varepsilon(-\omega)\,. \tag{2.71}$$

The representation of $\varepsilon(\omega)$ in the complex ω plane ensures that it is analytic on the upper complex plane ($\operatorname{Im}\omega \geq 0$), apart from a possible pole at $\omega = 0$. Indeed, for complex frequencies $z = x + iy$,

$$\varepsilon(z) = 1 + \chi_e(z) = 1 + \int_0^\infty d\tau\, \chi_e(\tau) e^{ix\tau} e^{-y\tau} \tag{2.72}$$

is analytic for all $\tau > 0$, including the upper limit $\tau \to \infty$, provided that $\chi_e(\tau \to \infty) \to 0$. The latter condition is met in dielectrics, but not in metals, where $\chi_e(\tau \to \infty) \to i\sigma/(\varepsilon_0\omega)$, and $\varepsilon(\omega)$ has a simple pole at $\omega = 0$. The analyticity of $\varepsilon(\omega)$ in the upper complex ω plane allows to employ Cauchy's theorem[11] on the expression

$$\oint_C dz \frac{\chi_e(z)}{z-\omega} = 0 \tag{2.73}$$

inside an appropriate closed contour C. After some straightforward algebra we obtain the Kramers-Kronig relations:

[11] Cauchy's theorem states that if a function $f(z)$ is complex analytic in a simply connected domain Ω, then $\int_C dz f(z) = 0$ for any simply closed contour C in Ω.

Box 2.9 | Kramers-Kronig relations

$$\operatorname{Re}\varepsilon(\omega) = 1 + \frac{2}{\pi}\mathscr{P}\int_0^\infty d\omega' \frac{\omega' \operatorname{Im}\varepsilon(\omega')}{\omega'^2 - \omega^2}, \tag{2.74a}$$

$$\operatorname{Im}\varepsilon(\omega) = \frac{\sigma_0}{\varepsilon_0\omega} - \frac{2\omega}{\pi}\mathscr{P}\int_0^\infty d\omega' \frac{\omega' \operatorname{Re}\varepsilon(\omega') - 1}{\omega'^2 - \omega^2}, \tag{2.74b}$$

where $\mathscr{P}$ denotes the principal value of the integral. As a consequence of these relations, the knowledge of the imaginary part of the dielectric function, for instance from absorption studies, can give information about the real part, and vice versa. At the same time, the Kramers-Kronig relations reveal a strict connection between dispersion and loss in a medium; strong dispersion is always accompanied by absorption.

2.3 Light–Matter Interactions

In this section we seek to answer a seemingly simple question; what happens when light encounters an object of size comparable to its wavelength? In Sect. 2.1 we established the theoretical tools required for the description of light, while in Sect. 2.2 we discussed how electrons in different media react when subjected to an EM field, so we are now fully equipped to delve into light–matter interactions. In this section we are mostly concerned with the interaction of light with either planar interfaces or spherical NPs that respond linearly to external perturbations.

Let us assume a finite object of arbitrary shape, illuminated by an EM wave as illustrated in Fig. 2.3 (for the example of a sphere). The electrons are set into oscillatory motion, following the oscillations of the incident electric field, while the nuclei, being much heavier, remain relatively motionless. The accelerated electric charges radiate EM energy in all directions. This combined excitation and reradiation process is called *scattering*. Furthermore, the excited electrons may transform part of their energy into other forms, most commonly to thermal energy; we call this process *absorption*. Therefore, part of the energy of the incident EM wave is transferred to the object and channeled to these two processes, that we collectively call *extinction*.

If we instead consider an object that extends infinitely along two dimensions and is finite in the third direction (a planar interface), the concept of scattering to all directions becomes obscure. Rather than scattering, the relative quantities in this case are *reflection* and *transmission*. The former corresponds to the process of the wave being deflected by the object, while the latter describes the process of the wave passing through to the other side of the boundary.

Nature offers an abundance of platforms with which one may study light–matter interactions. Different materials of various shapes and configurations, illuminated by different kinds of EM waves, can give rise to distinct phenomena, of which only

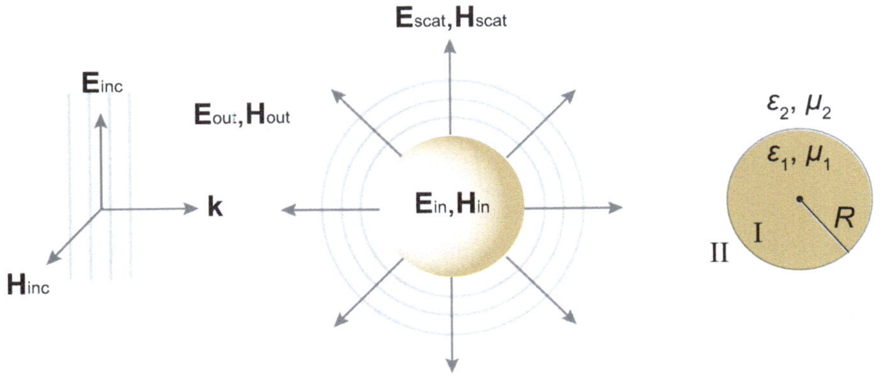

Fig. 2.3 **Mie theory**. A spherical NP of radius R and optical parameters (ε_1, μ_1) is illuminated by a plane EM wave $(\mathbf{E}_{inc}, \mathbf{H}_{inc})$. The incident wave is scattered in the form of outgoing spherical waves $(\mathbf{E}_{scat}, \mathbf{H}_{scat})$. In region II surrounding the NP (ε_2, μ_2), the field consists of the incident and the scattered ones. In region I inside the NP, the field $(\mathbf{E}_{in}, \mathbf{H}_{in})$ takes the form of standing spherical waves

a few can be discussed in the scope of a thesis. Of course, systems of high complexity require powerful computational resources and time-consuming investment in data analysis. However, one may reassuringly find that fundamental physics can be understood by studying the simplest systems. Mie theory of absorption and scattering of light by spherical NPs unarguably serves as an example of this.

2.3.1 Mie Theory

More than a century ago (1908), Gustav Mie developed a theory—but not necessarily the first one—regarding the absorption and scattering of light by spherical scatterers with arbitrary radius and refractive index, in an effort to understand the varied colors of small colloidal gold NPs suspended in water. Mie theory provides an accurate description of collective optical excitations in spherical NPs via closed-form and relatively simple analytic formulas. Most importantly, its basic principles serve as a stepping stone for the study of the optical response of spherical scatterers excited by any kind of EM waves. Here, we provide the key steps and conclusions of the theory, while a detailed derivation can be found in Appendix D, as well as in most textbooks on CED [1, 14–16].

Consider the system depicted in Fig. 2.3: a spherical NP of radius R, made up of a simple isotropic material with optical parameters (ε_1, μ_1), and excited by a plane EM wave $(\mathbf{E}_{inc}, \mathbf{H}_{inc})$ of wavelength comparable to its size. We wish to find the electric and magnetic field everywhere in space, i.e. the fields inside the NP $(\mathbf{E}_{in}, \mathbf{H}_{in})$, and the ones in the surrounding space (ε_2, μ_2), denoted in Fig. 2.3 as regions I and II, accordingly. The field in region II is the sum of the incident and the scattered field ($\mathbf{E}_{\text{scat}}, \mathbf{H}_{\text{scat}}$). The fields satisfy the Helmholtz vector wave equation (2.24), that admits incoming and outgoing spherical waves as solutions, expressed by spherical Bessel-type functions. The fields in the two regions can thus be decomposed into transverse electric (TE) and transverse magnetic (TM) waves of the form [1]

$$\mathbf{E}_{\text{in}}(\mathbf{r}) = \sum_{\ell m} \left\{ b^{\text{I}}_{\ell m} j_\ell(k_1 r)\mathbf{X}_{\ell m}(\hat{\mathbf{r}}) + \frac{i}{k_1} a^{\text{I}}_{\ell m} \nabla \times j_\ell(k_1 r)\mathbf{X}_{\ell m}(\hat{\mathbf{r}}) \right\}, \tag{2.75a}$$

$$\mathbf{H}_{\text{in}}(\mathbf{r}) = \frac{1}{Z_1} \sum_{\ell m} \left\{ a^{\text{I}}_{\ell m} j_\ell(k_1 r)\mathbf{X}_{\ell m}(\hat{\mathbf{r}}) - \frac{i}{k_1} b^{\text{I}}_{\ell m} \nabla \times j_\ell(k_1 r)\mathbf{X}_{\ell m}(\hat{\mathbf{r}}) \right\}, \tag{2.75b}$$

$$\mathbf{E}_{\text{inc}}(\mathbf{r}) = \sum_{\ell m} \left\{ b^{0}_{\ell m} j_\ell(k_2 r)\mathbf{X}_{\ell m}(\hat{\mathbf{r}}) + \frac{i}{k_2} a^{0}_{\ell m} \nabla \times j_\ell(k_2 r)\mathbf{X}_{\ell m}(\hat{\mathbf{r}}) \right\}, \tag{2.75c}$$

$$\mathbf{H}_{\text{inc}}(\mathbf{r}) = \frac{1}{Z_2} \sum_{\ell m} \left\{ a^{0}_{\ell m} j_\ell(k_2 r)\mathbf{X}_{\ell m}(\hat{\mathbf{r}}) - \frac{i}{k_2} b^{0}_{\ell m} \nabla \times j_\ell(k_2 r)\mathbf{X}_{\ell m}(\hat{\mathbf{r}}) \right\}, \tag{2.75d}$$

$$\mathbf{E}_{\text{scat}}(\mathbf{r}) = \sum_{\ell m} \left\{ b^{\text{II}}_{\ell m} h^+_\ell(k_2 r)\mathbf{X}_{\ell m}(\hat{\mathbf{r}}) + \frac{i}{k_2} a^{\text{II}}_{\ell m} \nabla \times h^+_\ell(k_2 r)\mathbf{X}_{\ell m}(\hat{\mathbf{r}}) \right\}, \tag{2.75e}$$

$$\mathbf{H}_{\text{scat}}(\mathbf{r}) = \frac{1}{Z_2} \sum_{\ell m} \left\{ a^{\text{II}}_{\ell m} h^+_\ell(k_2 r)\mathbf{X}_{\ell m}(\hat{\mathbf{r}}) - \frac{i}{k_2} b^{\text{II}}_{\ell m} \nabla \times h^+_\ell(k_2 r)\mathbf{X}_{\ell m}(\hat{\mathbf{r}}) \right\}, \tag{2.75f}$$

where ℓ and m are the orbital angular momentum quantum numbers that acquire integer values $\ell = 1, 2, 3, \ldots$ and $m = -\ell, -\ell + 1, \ldots, \ell - 1, \ell$. In Eqs. (2.75) in medium i, $Z_i = \sqrt{\mu_i \mu_0 / (\varepsilon_i \varepsilon_0)}$ stands for the impedance, $k_i = \sqrt{\varepsilon_i \mu_i}\omega / c$ is the wavenumber, j_ℓ is the spherical Bessel function of the first kind describing standing waves, h^+_ℓ the spherical Hankel function of the first kind describing outgoing spherical waves (see Appendix B) and $\mathbf{X}_{\ell m}(\hat{\mathbf{r}})$ are the vector spherical harmonics for an angle $\hat{\mathbf{r}} = (\theta, \phi)$ (see Appendix C).

In summary, the fields can be interpreted as a superposition of individual contributions produced by electric and magnetic dipoles ($\ell = 1$), quadrupoles ($\ell = 2$), octupoles ($\ell = 3$), etc., often referred to as *normal modes*, each weighted by a corresponding coefficient. In Appendix D we show that the expansion coefficients $a^{\text{I/II}}_{\ell m}$ and $b^{\text{I/II}}_{\ell m}$ can be determined by imposing the boundary conditions that dictate continuity of the tangential components of $\mathbf{E}$ and $\mathbf{H}$, yielding

$$a^{\text{II}}_{\ell m} = T_{\text{E}\ell} a^{0}_{\ell m}, \quad b^{\text{II}}_{\ell m} = T_{\text{M}\ell} b^{0}_{\ell m}, \tag{2.76a}$$

$$a^{\text{I}}_{\ell m} = T^{21}_{\text{E}\ell} a^{0}_{\ell m}, \quad b^{\text{I}}_{\ell m} = T^{21}_{\text{M}\ell} b^{0}_{\ell m}. \tag{2.76b}$$

The expansion coefficients of the incident wave $a^{0}_{\ell m}$ and $b^{0}_{\ell m}$ are given by Eqs. (D.14). The Mie coefficients connecting the scattered fields with the incident read

Box 2.10 | Mie coefficients of the scattered fields

$$T_{\mathrm{E}\ell} = \frac{\varepsilon_1 j_\ell(k_1 R)\Psi'_\ell(k_2 R) - \varepsilon_2 \Psi'_\ell(k_1 R) j_\ell(k_2 R)}{\varepsilon_2 h^+_\ell(k_2 R)\Psi'_\ell(k_1 R) - \varepsilon_1 \xi'_\ell(k_2 R) j_\ell(k_1 R)}, \tag{2.77a}$$

$$T_{\mathrm{M}\ell} = \frac{\mu_1 j_\ell(k_1 R)\Psi'_\ell(k_2 R) - \mu_2 \Psi'_\ell(k_1 R) j_\ell(k_2 R)}{\mu_2 h^+_\ell(k_2 R)\Psi'_\ell(k_1 R) - \mu_1 \xi'_\ell(k_2 R) j_\ell(k_1 R)}, \tag{2.77b}$$

where we introduced the Riccati-Bessel functions $\Psi_\ell(x) = x j_\ell(x)$ and $\xi_\ell(x) = x h^+_\ell(x)$, and the prime denotes the derivative with respect to the argument x evaluated at $x_i = k_i R$. For the Mie coefficients connecting the internal with the scattered fields, $T^{21}_{\mathrm{E/H}\ell}$, we refer the reader to Eqs. (D.19).

Using the Mie coefficients (2.77) alone, one can immediately study the absorption and scattering properties of a scatterer, quantified by the scattering, absorption and extinction cross sections, as

Box 2.11 | Optical cross sections

$$\sigma_{\mathrm{scat}} = \frac{2}{(k_2 R)^2} \sum_\ell (2\ell + 1)\Big\{|T_{\mathrm{E}\ell}|^2 + |T_{\mathrm{M}\ell}|^2\Big\}, \tag{2.78a}$$

$$\sigma_{\mathrm{ext}} = -\frac{2}{(k_2 R)^2} \sum_\ell (2\ell + 1)\mathrm{Re}\Big\{T_{\mathrm{E}\ell} + T_{\mathrm{M}\ell}\Big\}, \tag{2.78b}$$

$$\sigma_{\mathrm{abs}} = \sigma_{\mathrm{ext}} - \sigma_{\mathrm{scat}}. \tag{2.78c}$$

We note that the optical cross sections of Eqs. (2.78) are normalized to the geometrical cross section πR^2, and are hence dimensionless. The series in Eqs. (2.78) can be truncated at a certain value ℓ_{max} for which the summation over ℓ converges.

Albeit seemingly simple, one can gain a lot of information via these concise expressions. The Mie coefficients exhibit poles at certain frequency values ω_ℓ of the incident wave, different for each multipole; at ω_ℓ the optical cross sections reach local maxima. In practice, the corresponding spectra consist of a collection of Lorentzian curves centered around those resonant frequencies, at which the optical response of the NP is particularly efficient. We can make a rough estimation of these frequencies in the quasistatic limit.

Quasistatic Approximation The term *quasistatic* refers to the regime for which the finite speed of light is neglected, and the fields are regarded as propagating instantaneously. If the structure under study is small enough compared to the EM wavelength associated with the field fluctuation, then the electric field can be treated as constant in space, and only varying in time.[12] Using the limiting forms of Eqs. (B.7), one finds in the quasistatic limit ($k_i R \ll 1$)

[12] The prefix *quasi* refers to the fact that we still consider a frequency-dependent dielectric function $\varepsilon(\omega)$, which entails a time-dependent permittivity.

$$T_{E\ell} = \frac{i(\ell+1)(k_2R)^{2\ell+1}}{(2\ell+1)!!(2\ell-1)!!}\frac{\varepsilon_1-\varepsilon_2}{\ell\varepsilon_1+(\ell+1)\varepsilon_2}, \tag{2.79a}$$

$$T_{M\ell} = \frac{i(\ell+1)(k_2R)^{2\ell+1}}{(2\ell+1)!!(2\ell-1)!!}\frac{\mu_1-\mu_2}{\ell\mu_1+(\ell+1)\mu_2}. \tag{2.79b}$$

From the poles of the above equations, we deduce that small NPs exhibit maxima in their scattering, absorption, and extinction cross sections when the condition

$$\ell\varepsilon_1 + (\ell+1)\varepsilon_2 = 0 \tag{2.80}$$

is met. By introducing the Drude dielectric function (2.56) we may expect multipoles to arise at frequencies

$$\omega_\ell = \frac{\omega_p}{\sqrt{(1+1/\ell)\varepsilon_2+\varepsilon_\infty}}. \tag{2.81}$$

Assuming $\mu_1 = \mu_2 = 1$, only electric multipoles contribute to scattering and absorption ($T_{M\ell} = 0$). If, in addition, the NP is on the order of a few tens of nanometers, its optical response is adequately described by that of a point dipole, and only the electric dipolar mode is non-negligible. In particular, for $\ell = 1$ we find $T_{E1} = 2ik_2^3\alpha/3$, where

$$\alpha = R^3\frac{\varepsilon_1-\varepsilon_2}{\varepsilon_1+2\varepsilon_2} \tag{2.82}$$

is the dipolar polarizability of the NP [17].

Retardation With increasing NP size, the arguments adopted in the quasistatic limit are no longer a valid approximation. Given that the fields propagate with a finite speed c, the electric field propagates with *retardation*, which means that it reaches different points in space at different times. In this case, electric and magnetic loops can potentially be formed within the structure, giving rise to non-negligible magnetic modes and higher order TE and TM modes.

2.3.2 Fundamentals of Classical Plasmonics

A large part of this thesis is devoted to the field of plasmonics that studies the interaction of EM fields with the collective motion of free electrons in metals. Already in the development of the Drude model in Sect. 2.2.2.1, we saw that oscillating fields trigger plasma oscillations by setting the free electrons in a collective oscillatory motion. The quasiparticle of plasma oscillations called *plasmon* was initially proposed in 1952 by David Pines and David Bohm [18]. In the bulk of a metal, the plasmon oscillates at the plasma frequency ω_p, which is characteristic of the material. The elementary building block arising from the interaction of plasmons with light is called *plasmon polariton*, and, in terms of quasiparticle formulation, it can be interpreted as a plasmon coupled to a photon.

Different kinds of plasmons and plasmon polaritons with distinct properties can be encountered in different configurations. We discuss three basic types that constitute the foundation of plasmonics; the *bulk plasmon* (BP), the *surface plasmon*, and the *localized surface plasmon* (LSP). We describe plasmons as emerging from the classical Maxwell equations, and consider structures large enough so that LRA is a valid approximation. We thus lay the foundation of classical plasmonics, while also establishing the reference point to which mesoscopic plasmonics in the following sections is juxtaposed.

2.3.2.1 Bulk Plasmons

Let us study the propagation of waves inside a metal that is infinite in all directions, behaving as a free electron gas (FEG). Once illuminated, TE waves propagate inside the material following the Helmholtz equation (2.24) for a wavenumber k. The system can be adequately described by a simple Drude model $\varepsilon = 1 - (\omega_\mathrm{p}/\omega)^2$ and relative permeability $\mu = 1$. Combined with $k = \sqrt{\varepsilon}\omega/c$ this yields the dispersion relation of the FEG

Box 2.12 | FEG and transverse BPP dispersion relation

$$\omega = \sqrt{\omega_\mathrm{p}^2 + c^2k^2}\ . \tag{2.83}$$

Plasma oscillations that follow the dispersion relation of the free electron gas are transverse *bulk plasmon polaritons* (BPPs). One can observe in Fig. 2.4b (green solid line) that the BPP modes exist above the plasma frequency ω_p, at which the Drude material has a positive permittivity and behaves like a dielectric.

According to relation (2.83), at large wavelengths ($k \to 0$), the frequency does not tend to zero, like it would for EM waves in free space, but to the bulk plasmon frequency

$$\omega \xrightarrow[k\to 0]{} \omega_\mathrm{b} = \omega_\mathrm{p}\ . \tag{2.84}$$

Since, at this frequency, the Drude permittivity goes to zero, longitudinal waves can develop inside the metal [see Eq. (2.35)]. Those are the longitudinal bulk plasmons, described by the following dispersion relation

Box 2.13 | Longitudinal BP dispersion relation

$$\varepsilon(\omega, \mathbf{k}) = 0\ . \tag{2.85}$$

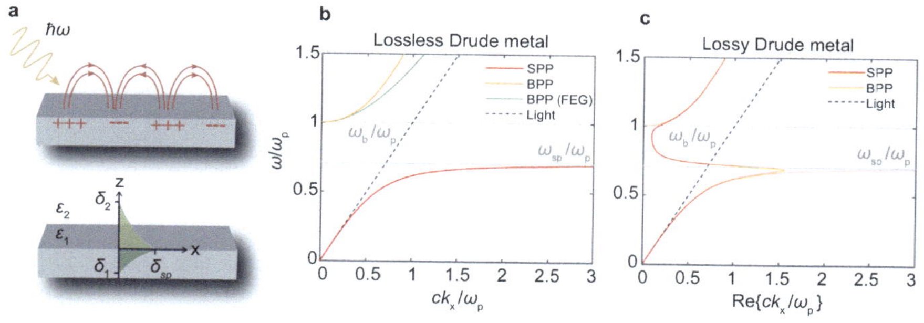

Fig. 2.4 SPP and BPP dispersion relation. **a** Schematic illustration of SPPs generated at the flat interface between air (ε_2, $\mu_2 = 1$) and a metal ($\mu_1 = 1$) under illumination. **b** The dispersion relation of the plasmon modes sustained in a lossless Drude metal ($\gamma = 0$). **c** Same as in **b**, here for a lossy Drude metal with $\gamma = 0.05\omega_p$. The red solid line corresponds to the dispersion of SPPs, while the yellow line shows the BPPs according to Eq. (2.90). The green solid line depicts the dispersion relation of BPPs within the FEG model (i.e. we disregard the interface) as given by Eq. (2.83), and the black dashed line shows the dispersion relation of light in free, space $c = \omega/k_x$. The SP frequency, ω_{sp}, and the BP frequency, ω_b are denoted with gray solid lines. In panel **c** the faded red and yellow lines trace the SPP and BPP curves of panel **b**

Notice that we have included the dependence on **k**, because the dispersion relation for these plasmons is closely related to spatial dispersion. We return to this point with more details in Sect. 2.3.5.

2.3.2.2 Surface Plasmons

SPs are propagating waves of the electron density confined to a flat dielectric–metal interface. They were first described by Ritchie [19] (1957) in spectra of thin foils excited by fast electrons, and later confirmed experimentally by Powell and Swan [20]. Not long after that, followed the first observation of SPPs excited by electron beams, by Teng and Stern [21].

Let us outline here the main findings of the theory of SPPs.[13] Consider the system depicted in Fig. 2.4a; medium 1 is a metal with permittivity $\varepsilon_1 < 0$, and medium 2 is dielectric, with $\varepsilon_2 > 0$; for the sake of simplicity, we disregard losses and set $\mu_1 = \mu_2 = 1$. The application of an external electric field leads to a separation of charges at the surface of the metal and, eventually, to restoring forces that sustain the SP oscillations. Suppose that we illuminate the system with an EM field characterized by wavevector $k_2 = (k_x, 0, k_{2z})$, where k_x and k_{2z} are real, so that $k_2^2 = \varepsilon_2(\omega/c)^2 > 0$. Then the z component of the wavevector in the metal is determined from the dispersion relation

$$k_{1z}^2 = k_1^2 - k_{1x}^2 = \frac{\omega^2}{c^2}\varepsilon_1 - k_x^2\,, \tag{2.86}$$

[13] More detailed derivations can be found in numerous textbooks, see, for example, Refs. [14, 22].

where $k_{1x} = k_x$ stems from the conservation of the wavevector parallel to the interface. Since $\varepsilon_1 < 0$, we deduce that $k_{1z}^2 < 0$. We can then write $\mathbf{k}_i = k_x\hat{\mathbf{x}} + i\kappa_i\hat{\mathbf{z}}$ (with $i = 1, 2$ being the medium labels), where k_x, as discussed in Sect. 2.1.6, is related to the wave propagation, and $\kappa_i^2 = k_x^2 - \varepsilon_i(\omega/c)^2$ to the wave attenuation.

The wave equation for this system admits solutions of the form of TM waves,[14] which, assuming harmonic time-dependence, read

$$\mathbf{E}_2(\mathbf{r}) = [E_{2x}\hat{\mathbf{x}} + E_{2z}\hat{\mathbf{z}}]e^{ik_x x}e^{-\kappa_2 z}\,, \tag{2.87a}$$

$$\mathbf{H}_2(\mathbf{r}) = H_{2y}e^{ik_x x}e^{-\kappa_2 z}\hat{\mathbf{y}} \tag{2.87b}$$

in the dielectric medium, and

$$\mathbf{E}_1(\mathbf{r}) = [E_{1x}\hat{\mathbf{x}} + E_{1z}\hat{\mathbf{z}}]e^{ik_x x}e^{\kappa_1 z}\,, \tag{2.88a}$$

$$\mathbf{H}_1(\mathbf{r}) = H_{1y}e^{ik_x x}e^{\kappa_1 z}\hat{\mathbf{y}} \tag{2.88b}$$

in the metal. The above expressions explicitly show that the wave is confined to the interface, with a confinement length in the dielectric material of the order of $\delta_2 = 1/|k_{2z}|$, and skin depth into the metal of the order of $\delta_1 = 1/|k_{1z}|$. For instance, for $\lambda = 450$ nm-light incident on an air–silver interface, the confinement length is of the order of $\delta_2 = 200$ nm, and the skin depth is about $\delta_2 = 15$ nm [22].

By invoking Maxwell's equations, as well as the continuity of the tangential electric fields, one can determine a condition for the excitation of SPPs

$$\frac{\varepsilon_2}{\kappa_2} + \frac{\varepsilon_1}{\kappa_1} = 0\,. \tag{2.89}$$

To satisfy this expression for real and positive κ_i, the signs of ε_1 and ε_2 must be opposite, which is automatically fulfilled for the dielectric–metal interface for typical metals in the visible and near-infrared spectrum. Introducing Eq. (2.86) into Eq. (2.89) one obtains the SPP dispersion relation.

Box 2.14 | SPP dispersion relation and conditions

The SPP dispersion relation reads

$$k_x = \frac{\omega}{c}\sqrt{\frac{\varepsilon_1\varepsilon_2}{\varepsilon_1 + \varepsilon_2}}\,, \tag{2.90}$$

provided that the following conditions hold

$$\varepsilon_1\varepsilon_2 < 0 \quad \text{and} \quad \varepsilon_1 + \varepsilon_2 < 0\,. \tag{2.91}$$

[14] In conventional metals SPPs cannot be excited by TE wave polarization [14].

The second inequality in Box 2.14 emerges from the requirement to have a real wavenumber k_x.

Equation (2.90) yields two solutions, one above and one below ω_p, as presented in Fig. 2.4b. Only the latter fulfils conditions (2.91) and, therefore, describes an SPP mode bound at the flat interface between air and the lossless metal. The plasmon polariton lying above ω_p is a type of propagating BPP,[15] like the one encountered in the FEG, but with a slightly different asymptotic behavior for increasing wavenumber k_x; in the FEG the branch follows $\omega \to ck_x$ (see green line in Fig. 2.4b), while in the dielectric–metal system the existence of a second material manifests through its dielectric function entering the asymptotic limit as $\omega \to ck_x\sqrt{(\varepsilon_2 + 1)/\varepsilon_2}$ (see yellow line in Fig. 2.4b). Regarding the SPP mode, according to Fig. 2.4b, at low wavevectors ($k_x \to 0$) it tends to the photon dispersion relation of light in free space (see black dashed line), meaning that it features mainly photon-like characteristics and poor localization. In contrast, for $k_x \to \infty$ the frequency of the SPP mode approaches asymptotically the surface plasmon frequency $\omega_{sp} = \omega_p/\sqrt{\varepsilon_\infty + \varepsilon_2}$. Assuming $\varepsilon_\infty = \varepsilon_2 = 1$, that frequency is equivalent to

$$\omega \xrightarrow[k \to \infty]{} \omega_{sp} = \frac{\omega_p}{\sqrt{2}} . \tag{2.92}$$

This asymptotic trend showcases the argument of subwavelength confinement achieved by plasmons; the wavelength of the SPP mode can be significantly shorter than the free-space wavelength of light. However, since the SPP mode always lies below the photon dispersion relation of free-space light, it cannot couple to it due to momentum mismatch. One can, however, implement other coupling schemes like the Otto and Kretschmann configurations to circumvent the mismatch [22].

In reality all metals are accompanied by non-negligible losses, thus changing the ideal picture painted in Fig. 2.4b. Once losses are introduced to the Drude model, k_x assumes an imaginary part that translates to damping of the propagating plasmon, and also determines the propagation length $\delta_{sp} = 1/\mathrm{Im}(k_x)$. In this case, as seen in Fig. 2.4c, the dispersion relation allows for a finite maximum value k_x for the SSP mode, and then exhibits a backbend around ω_{sp} (where k_x is complex) that connects the SPP with the BPP branch.

2.3.2.3 Localized Surface Plasmons

LSPs are standing waves of the electron density confined on the surface of finite metallic structures, such as a disk or a sphere. As in the case of SPPs at infinite planar interfaces, the illumination of a spherical metallic surface initiates a self-sustained oscillation of free electrons [see Fig. 2.5a], owing to restoring forces. The

[15] Notice that the original assumption $\varepsilon_1 < 0$ does not hold anymore for the Drude metal above ω_p. As a consequence, k_{1z} is a real number and the plasmon polariton propagates along the z direction in the bulk of the material.

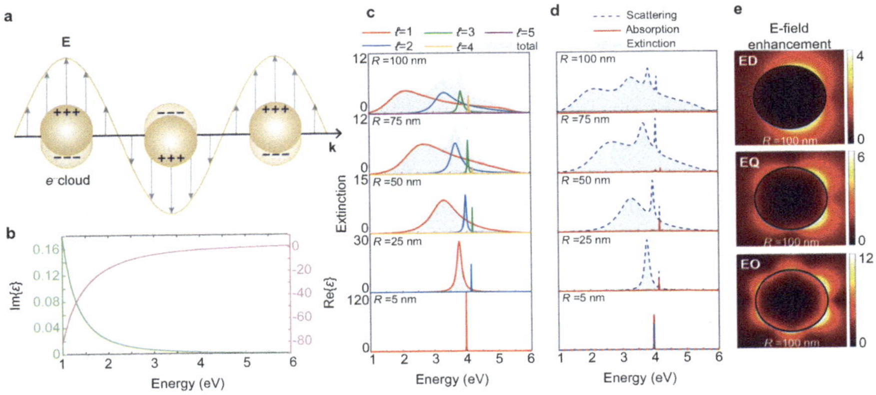

Fig. 2.5 LSPs in spherical metallic NPs. **a** Schematic illustration of the formation of LSPs on the surface of a metallic sphere illuminated by a transverse EM wave. **b** Real (right y axis) and imaginary part (left y axis) of the Drude dielectric function of Eq. (2.56) with $\varepsilon_\infty = 3.3$, $\hbar\omega_{\mathrm{p}} = 9.17$ eV and $\hbar\gamma = 21$ meV, used for all spectra presented in this figure. **c** Normalized extinction cross section (gray shaded area) and modal decomposition into the first five electric multipoles (red, blue, green, yellow and purple solid lines) for metallic NPs of increasing radii $R = 5, 25, 50, 75, 100$ nm from bottom to top. **d** Normalized extinction (gray shaded area), absorption (dark red solid line) and scattering (dark blue dashed line) cross sections for the same NPs as in panel **c**. **e** Electric field enhancement, given by the norm of the induced field over the incident field, calculated for the $R = 100$ nm sphere at the energies of the ED, EQ, and EO mode. For all three panels we consider an incident plane wave propagating along the horizontal axis with polarization parallel to the perpendicular axis, as illustrated in panel **a**

finiteness of the structure constrains the induced fields to develop as standing waves on the curved interface, since propagation is not possible. In this thesis, LSPs are the type of plasmons that concerns us the most. Having established Mie theory in Sect. 2.3.1, we may now utilize our results to describe the excitation of LSPs in spherical metallic NPs.

Consider the NP depicted in Fig. 2.5a, described by the dielectric function of Fig. 2.5b, i.e. a Drude permittivity for silver, with $\varepsilon_\infty = 3.3$, $\hbar\omega_{\mathrm{p}} = 9.17$ eV and $\hbar\gamma = 21$ meV in Eq. (2.56). In Fig. 2.5c we present the extinction cross section for various sphere sizes (5 nm $< R <$ 100 nm), as calculated by Eq. (D.26c), as well as the corresponding multipole decomposition. Only electric modes contribute to the extinction, while the magnetic ones are negligible, apart from a very broad low-intensity shoulder at around 5 eV in the largest sphere of radius $R = 100$ nm. According to Eq. (2.81), in the quasistatic approximation the dipolar mode is expected at around $\hbar\omega_1 = 3.98$ eV, but clearly, the prediction is accurate only for the smallest sphere of radius $R = 5$ nm. Larger NPs can accommodate higher order multipoles, and they increasingly deviate from the quasistatic approximation; the spectra redshift due to increased retardation and the subsequent suppression of the driving forces acting on the free charge carriers. Moreover, we observe that same-order modes exhibit higher radiation damping as the size of the NP increases. Figure 2.5d reveals that this occurs due to the fact that larger spheres radiate light more effectively, whereas in smaller NPs the main channel of dissipation is absorption. This argument holds mainly for the dipolar mode, since higher order modes, i.e. quadrupoles, octupoles, etc. do not radiate that effectively. According to Eq. (2.85), the longitudinal BP mode should emerge at the frequency where the real part of the permittivity becomes zero. That is not the case here, since, in traditional Mie theory, we only allow for the existence of transverse waves inside the material.

Finally, in Fig. 2.5e we show the electric field enhancement for the largest sphere evaluated at the energies at which the three first modes exhibit a peak, more specifically at the electric dipolar mode (ED), the quadrupolar (EQ), and the octupolar (EO) mode, at $\hbar\omega_1 = 2.14$ eV, $\hbar\omega_2 = 3.32$ eV, and $\hbar\omega_3 = 3.64$ eV, respectively; indeed the field profiles resemble the electric field of a point ED, EQ and EO accordingly, albeit with a slight asymmetry owing to the considerable overlap of modes of different orders. The field enhancement and confinement on the NP surface, as well as the suppression of the field inside the metal, is evident. For the dipolar mode the confinement is of the order of $\lambda = 58$ nm for $\lambda = 580$ nm incident light, providing, thus, a rough tangible proof that subwavelength confinement is possible.

2.3.3 Mie Resonances in Dielectrics

In Sect. 2.2.2.2 we established that when a dielectric medium is subjected to an EM field, its bound electrons are displaced away from the ions. The interaction of dielectrics with light gives rise to a self-sustained oscillation of those bound electrons in the material. In contrast to metals, where the charge carriers move freely, and subsequently accumulate at the surface of a finite structure, in dielectrics the electrons are constrained by the ions to oscillate locally, thereby creating volume polarization charges. This feature, combined with the effect of retardation in large structures, allows dielectric NPs to generate current loops in their volume. Therefore, the EM fields scattered by spherical dielectric NPs, unlike metals, contain considerable contributions of magnetic multipoles, in addition to the electric ones.

We indeed see that in Fig. 2.6c, where we study the optical response of silicon (Si) NPs of different sizes, whose dielectric function is given by the experimental values found in Ref. [23] and plotted in Fig. 2.6b. Interestingly, the magnetic dipolar (MD) mode is the lowest-energy mode supported by the NP. One may also observe that each mode exhibits more than one peaks. These features correspond to higher-radial-order modes, and are responsible for the asymmetric profile of the Lorentzian curves. Once again, larger spheres host higher-order Mie resonances that redshift and broaden with increasing size. Above 3 eV the absorption and scattering spectra of Fig. 2.6d show a step-like contribution, stemming from interband transitions in Si, as inferred from the increased imaginary part of its dielectric function at these energies.

The origin of dielectric resonances whose permittivity does not assume negative values is not that obvious mathematically, but it can be described utilizing Padé

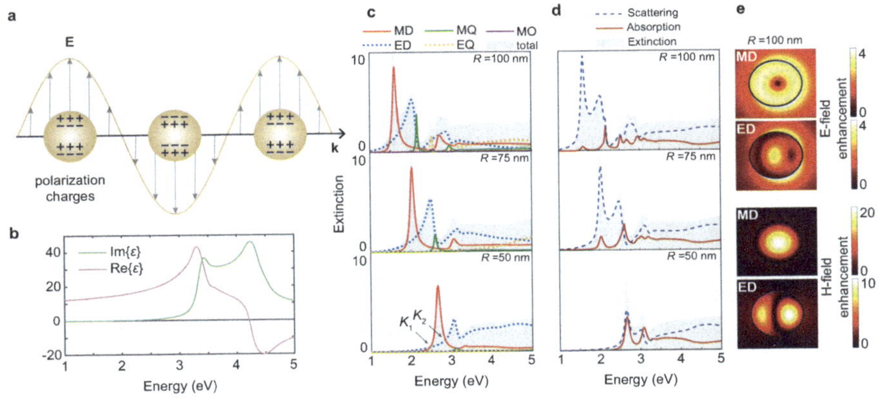

Fig. 2.6 Mie resonances in spherical dielectric NPs. a Schematic illustration of the formation of Mie resonances in a dielectric sphere illuminated by a transverse EM wave. **b** Real (pink curve) and imaginary part (green curve) of the dielectric function of Si according to Ref. [23], used for all spectra presented in this figure. **c** Normalized extinction cross section (gray shaded area) and modal decomposition into the MD, magnetic quadrupolar (MQ), and magnetic octupolar (MO) modes, with red, green, and purple solid lines respectively, and the ED and EQ mode with blue and yellow dotted lines respectively, for Si NPs of increasing radii $R = 50, 75, 100$ nm from bottom to top. **d** Normalized extinction (gray shaded area), absorption (dark red solid line) and scattering (dark blue dashed line) cross sections for the same NPs as in panel **c**. **e** Electric (top) and magnetic field enhancement (bottom), given by the norm of the induced field over the incident field, calculated for the $R = 100$ nm sphere at the energies of the MD and ED mode. For all four panels we consider an incident plane wave propagating along the horizontal axis with polarization parallel to the perpendicular axis, as illustrated in panel **a**

approximants[16] [24]. In our example, the [3/3] Padé approximant expansion for T_{E1} reads

$$T_{\mathrm{E1}} = -\frac{2i\alpha}{3R^3}\frac{x^3}{\left(1-\frac{3}{5}\frac{\varepsilon_1-2}{\varepsilon_1+2}x^2 - i\frac{2}{3}\frac{\varepsilon_1-1}{\varepsilon_1+2}x^3\right)}, \tag{2.93}$$

with $x = k_2 R$ and α the polarizability of Eq. (2.82), and exhibits a pole at

$$\varepsilon_1 = -2\frac{15i + 9ix^2 - 5x^3}{15i - 9ix^2 + 10x^3}. \tag{2.94}$$

For small x the pole emerges at

$$\varepsilon_1 = -2 - \frac{12}{5}x^2\left(1+\frac{3}{5}x^2\right) - 2ix^3\left(1+\frac{7}{5}x^2\right) + \ldots . \tag{2.95}$$

[16] The Padé approximant is a special type of rational approximation used particularly for many physical problems where a complex resonant physical quantity is observed. The approximant of order $[p/q]$ is constructed by expressing a function f as a fraction of two polynomials $f(x) = P(x)/Q(x)$ of orders p and q, respectively.

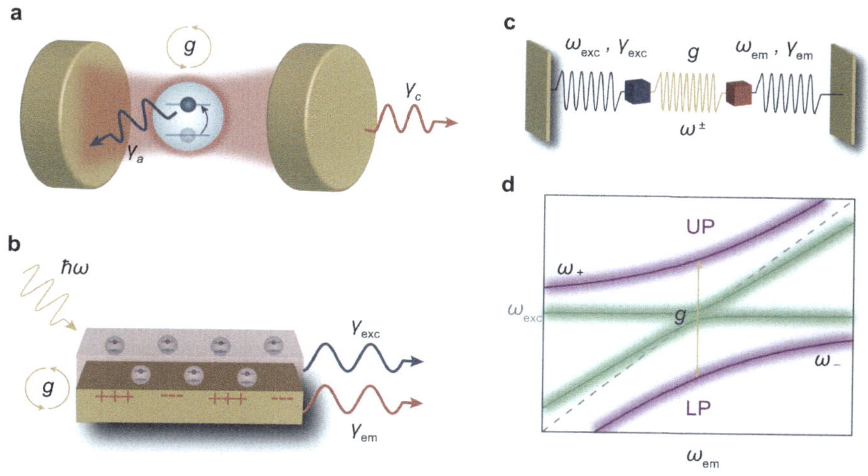

Fig. 2.7 **Strong light–matter interactions. a** Strong coupling between a two-level atom and a cavity. **b** Strong coupling between SPPs and excitons at the interface between a plasmonic and an excitonic medium. **c** The CHO model. **d** CHO dispersion relation in the strong (purple lines) and weak (green lines) coupling regime, where the gray dashed lines trace the modes of the uncoupled oscillators. The UP and LP branches and their respective frequencies $\omega_{\pm}$ emerge from the system of Eqs. (2.98) for a given exciton transition frequency ω_{exc} and varying light frequency ω_{em}. The CHO are strongly coupled when the avoided crossing between the polaritonic branches is larger than the linewidth of the individual modes, represented here by the purple and green faded lines

As expected, this term reproduces the pole $\varepsilon_1 = -2\varepsilon_2$ of the quasistatic approximation, that can only be fulfilled by metals. Beyond that limit, we can observe that condition (2.95) is met at complex x; even in lossless NPs, the Mie resonances appear at complex frequencies due to the fact that the system is an open resonator that suffers radiation damping. The real and imaginary terms of Eq. (2.95) introduce size-dependent shifting of the resonant position, and damping, respectively.

Finally, in Fig. 2.6e we show the electric and magnetic field enhancement of the largest sphere evaluated at the energies of the MD, and ED modes. The field profiles showcase the excitation of polarization charges within the volume of the NP. Moreover, we observe that the electric field enhancement in dielectric NPs is lower compared to their plasmonic counterparts, but they offer substancial magnetic field enhancement. Unlike plasmonic nanospheres that are omnidirectional scatterers, dielectric ones do exhibit a preference in scattering direction. Characteristically, under certain conditions for the values of ε and μ (known as *Kerker conditions*), the induced electric and magnetic dipole can oscillate in phase or out of phase, leading to zero-backward and zero-forward scattering respectively, with respect to the direction of the incident plane wave [25]. In the silicon NP presented in Fig. 2.6, zero backward scattering is observed at the lowest energy where the MD and ED mode intersect (first Kerker condition), and zero forward scattering occurs at the next MD-ED intersection point (second Kerker condition), as the labels $K_{1/2}$ in Fig. 2.6c suggest.

2.3.4 *Strong and Weak Light–Matter Interactions*

2.3.4.1 Strong Coupling

In the previous sections we discussed the physics behind three distinct kinds of EM modes that result from the interaction of light and matter: SPPs and LSPs in metals, and Mie resonances in dielectric NPs. All are resonant modes stemming from the coupling of light with the electrons of the polarizable medium. In reality, there are numerous other types of quasiparticles resulting from the coupling of photons with all sorts of polar excitations in materials, all of which we address with the term *polaritons*. In the special case of an EM mode coupling to an optical transition, the system may form a hybrid light–matter state. Inspired from the field of quantum optics, we call this process *strong coupling*.

The usual problem of light–matter interactions in quantum optics, also known as the *Rabi problem*, consists of a cavity and an atom, modeled as a two-level system, supporting a transition from the ground to an excited energy state [26], as depicted in Fig. 2.7a. On resonance, when the frequency of the photon in the cavity mode coincides with the frequency of the atomic transition, the photon is absorbed by the atom, and then released due to spontaneous emission, only to be absorbed again. This cycle of energy exchange between cavity and atom may occur multiple times, until the photon escapes the system, given that the cavity cannot be ideal and will thus allow energy dissipation.

Strong light–matter interactions in the field of photonics translate to the coupling of a resonator supporting EM modes with media that sustain optical transitions (see e.g. Fig. 2.7b), with the most commonly used being the excitonic materials. An *exciton* is a Coulomb-bound state of an electron–hole pair formed when a material with a bandgap, such as an insulator and or a semiconductor, absorbs light of energy higher than its bandgap.[17] The exciton itself can couple to light and form an *exciton polariton*. Good candidates for exciton generation are low dimensional semiconductor structures, like quantum dots[18] and transition-metal dichalcogenide (TMDC)[19] layers [29]. Besides semiconductors, aggregates of atoms or molecules (to ensure collective optical response), such as organic dyes and *J*-aggregates, can also sustain strong excitonic transitions [30].

Interestingly, once strong coupling between the resonator and the excitonic medium is achieved, the system exhibits normal modes of part-light and part-matter nature, as well as spatial coherence [31]. In order to understand the emergence of these light–matter hybrids, we may emulate the system with one that is better understood. Given the type of restoring forces developed in the media we study here upon

[17] In the electronic band structure of insulators and semiconductors, the bandgap refers to the energy separation between the top of the valence band and the bottom of the conduction band [8].

[18] Semiconductor NPs of a few nanometers in size [27].

[19] Class of materials of the type MX_2, with M a transition metal (Mo, W, etc.) and X a chalcogen atom (S, Se, or Te). Certain TMDCs have direct bandgaps, which makes them well-suited for optoelectronic applications [28].

illumination, the coupled harmonic oscillators (CHO) model seems the most sensible choice.

Coupled Harmonic Oscillators Model Let us consider the two harmonic oscillators illustrated in Fig. 2.7c, which are identified by the labels em and exc (standing for EM mode and exciton, respectively). They are characterized by their respective frequencies ω_{em} and ω_{exc}, decay rates γ_{em} and γ_{exc}, and unitary masses for simplicity. For incident light of frequency ω, the force driving each oscillator takes the form $F(t) = f_0 \exp(-i\omega t)$, and the equations of motion for the position x of each oscillator read

$$\frac{d^2 x_{\text{exc}}}{dt^2} + \gamma_{\text{exc}} \frac{dx_{\text{exc}}}{dt} + \omega_{\text{exc}}^2 x_{\text{exc}} - g \frac{dx_{\text{em}}}{dt} = F(t) , \tag{2.96a}$$

$$\frac{d^2 x_{\text{em}}}{dt^2} + \gamma_{\text{em}} \frac{dx_{\text{em}}}{dt} + \omega_{\text{em}}^2 x_{\text{em}} + g \frac{dx_{\text{exc}}}{dt} = F(t) , \tag{2.96b}$$

where g is the coupling parameter. The extinction of the excitonic systems studied in this thesis is typically significantly smaller than that of the resonators they couple to. Therefore, we assume that the extinction cross section of the system is proportional to the power, i.e. the work done on oscillator em by the force per unit time, as

$$\sigma_{\text{ext}} \propto \left\langle F(t) \frac{dx_{\text{em}}(t)}{dt} \right\rangle , \tag{2.97}$$

where the brackets denote the average over a period of the oscillation. The aforementioned assumption is equivalent to setting $F(t) = 0$ in Eq. (2.96a). At steady state, both oscillators follow the harmonic time dependence $x_j(t) = \tilde{x}_j \exp(-i\omega t)$, and the system of equations ultimately reduces to

$$\begin{pmatrix} \omega_{\text{exc}}^2 - \omega^2 - i\gamma_{\text{exc}}\omega & ig\omega \\ -ig\omega & \omega_{\text{em}}^2 - \omega^2 - i\gamma_{\text{em}}\omega \end{pmatrix} \begin{pmatrix} \tilde{x}_{\text{exc}} \\ \tilde{x}_{\text{em}} \end{pmatrix} = \begin{pmatrix} 0 \\ f_0 \end{pmatrix} , \tag{2.98}$$

with solutions[20]

$$\tilde{x}_{\text{em}} = \text{Re}\left\{ \frac{f_0(\omega_{\text{em}}^2 - \omega^2 - i\gamma_{\text{em}}\omega)}{(\omega_{\text{em}}^2 - \omega^2 - i\gamma_{\text{em}}\omega)(\omega_{\text{exc}}^2 - \omega^2 - i\gamma_{\text{exc}}\omega) - g^2\omega^2} \right\} , \tag{2.99a}$$

$$\tilde{x}_{\text{exc}} = \text{Re}\left\{ \frac{-if_0 g\omega}{(\omega_{\text{em}}^2 - \omega^2 - i\gamma_{\text{em}}\omega)(\omega_{\text{exc}}^2 - \omega^2 - i\gamma_{\text{exc}}\omega) - g^2\omega^2} \right\} . \tag{2.99b}$$

Inserting Eq. (2.99a) in Eq. (2.97) we obtain

---

[20] We pick the real part of the algebraic solution of Eqs. (2.98), since the displacement $\tilde{x}_j$ is a physical quantity and thus it must be real.

$$\sigma_{\text{ext}} \propto \omega \text{Im} \left\{ \frac{(\omega_{\text{exc}}^2 - \omega^2 - i\gamma_{\text{exc}}\omega)}{(\omega_{\text{em}}^2 - \omega^2 - i\gamma_{\text{em}}\omega)(\omega_{\text{exc}}^2 - \omega^2 - i\gamma_{\text{exc}}\omega) - g^2\omega^2} \right\} . \tag{2.100}$$

When the two systems are at resonance, $\omega_{\text{em}} = \omega_{\text{exc}} = \omega_0$, the above equation simplifies to

$$\sigma_{\text{ext}} \propto \omega \text{Im} \left\{ \frac{(\omega_0 - \omega - i\gamma_{\text{exc}}\omega/2)}{\omega_0\sqrt{4g^2 - (\gamma_{\text{em}} - \gamma_{\text{exc}})^2}} \left(\frac{1}{\omega - \omega_+} - \frac{1}{\omega - \omega_-} \right) \right\} , \tag{2.101}$$

where

$$\omega_\pm = \omega_0 + \frac{i(\gamma_{\text{em}} + \gamma_{\text{exc}})}{4} \pm \frac{1}{2}\sqrt{g^2 - \left(\frac{\gamma_{\text{em}} - \gamma_{\text{exc}}}{2}\right)^2} \tag{2.102}$$

are the normal modes of the coupled system.

In Fig. 2.7d we plot the extinction cross section of Eq. (2.100) for a certain value of the excitonic transition frequency ω_{exc}, while varying the photon frequency ω_{em}. We indeed observe the emergence of two normal modes—from this point on, we refer to the two branches as *lower* (LP) and *upper polariton* (UP)—that are different from the ones supported by the individual constituents, highlighted with dashed lines. At resonance they exhibit an avoided crossing or *anticrossing*, with an energy separation equal to $\omega_+ - \omega_- \approx g$, for $\gamma_{\text{exc}} = \gamma_{\text{em}} = 0$ in Eq. (2.102). Therefore, this energy separation, known as *mode splitting*, is a measure of the coupling strength g and a footprint of strong coupling. Taking into account that in a realistic system losses are non-negligible, strong coupling need be assessed as compared to a critical value determined by the damping rates of the two resonant systems individually. It is clear that one cannot speak of strong coupling if the rate of energy dissipation in either of the two constituents is faster than the rate of energy exchange.

A strict definition of the onset of strong coupling does not really exist. Characteristically, Törmä and Barnes note that the definition of strong coupling is context sensitive, depending on several conventions within the fields of physics involved [32]. As far as we are concerned, we follow the strong-coupling criterion found in said work, which aligns with the CHO model developed above; the system operates in the strong coupling regime if a mode splitting in scattering and absorption spectra (if there is substantial absorption to begin with) is discernible. For this condition to be met, the real part of energy separation [square root in Eq. (2.102)] must be larger than the widths of the new modes, yielding the inequality

$$\sqrt{g^2 - \left(\frac{\gamma_{\text{em}} - \gamma_{\text{exc}}}{2}\right)^2} > \frac{\gamma_{\text{em}} + \gamma_{\text{exc}}}{2} . \tag{2.103}$$

We therefore accept that a system is strongly coupled whenever the following criterion holds:

Box 2.15 | Strong coupling criterion

$$g > \sqrt{\frac{\gamma_{\mathrm{em}}^2 + \gamma_{\mathrm{exc}}^2}{2}}\,, \tag{2.104}$$

otherwise the system is *weakly* coupled.

2.3.4.2 Weak Coupling

Interesting phenomena arise even before entering the strong coupling regime. Here we examine the coupling between a spherical NP and an oscillating dipole, which captures quite accurately the various phenomena observed in weakly coupled systems.[21] The analysis done in this section constitutes the theoretical foundation for different kinds of spectroscopies, including SERS[22] and EELS; the latter is discussed thoroughly in Chap. 5, albeit in a CED framework. To adopt the terminology typically used in the field, we refer to the entity that sustains the optical transition as *quantum emitter* (QE), although our analysis is based to a large extent on CED. As we show in what follows, the QE experiences an enriched EM environment due to the presence of the scatterer, that modifies its excitation and emission properties [34].

Consider the system depicted in Fig. 2.8; a QE, modeled by an electric dipole moment $\mathbf{p}_{\mathrm{d}}$, is located at distance $\mathbf{r}_{\mathrm{d}}$ from the center of a spherical NP of radius R ($r_{\mathrm{d}} > R$), embedded in a non-absorbing host medium (ε_2, μ_2). The QE is driven by light of oscillation frequency ω, inducing a current distribution

$$\mathbf{J}(\mathbf{r}) = -i\omega\, \mathbf{p}_{\mathrm{d}}\, \delta(\mathbf{r} - \mathbf{r}_{\mathrm{d}})\,. \tag{2.105}$$

One can then relate the total electric field to the QE excitation via the Green's dyadic function, using Eq. (2.22), as

$$\mathbf{E}_{\mathrm{tot}}(\mathbf{r}) = \omega^2 \mu_2 \mu_0\, \overleftrightarrow{\mathscr{G}}_{\mathrm{tot}}(\mathbf{r}, \mathbf{r}_{\mathrm{d}}, \omega) \cdot \mathbf{p}_{\mathrm{d}}\,, \tag{2.106}$$

where $\overleftrightarrow{\mathscr{G}}_{\mathrm{tot}} = \overleftrightarrow{\mathscr{G}}_0 + \overleftrightarrow{\mathscr{G}}_{\mathrm{ref}}$ is the total Green's dyadic function that includes a free-space part $\overleftrightarrow{\mathscr{G}}_0$ as well as a reflected part $\overleftrightarrow{\mathscr{G}}_{\mathrm{ref}}$ due to the presence of a structure modifying the photonic environment.[23] Knowing the Green's function of the problem, in our

[21] It is worth mentioning that the analysis provided in this section is actually general; it describes the weak coupling between any kind of scatterer or non-trivial photonic environment with a QE. In the scope of this thesis, however, we are only interested in spherical NPs.

[22] Raman scattering refers to the inelastic scattering of photons by matter. The technique is very useful in the characterization of nanomaterials, but it is based on very weak effects. In SERS the Raman scattering cross-section is increased significantly by absorbing the molecules onto rough metal structures [33].

[23] We assume that the scatterer only affects the local environment but not the transitions of the QE.

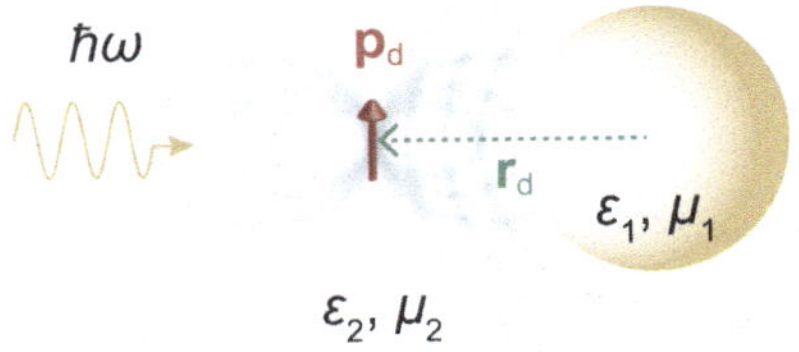

Fig. 2.8 Weak NP–QE interaction. A system consisting of a spherical NP with optical parameters (ε_1, μ_1) and a QE at position $\mathbf{r}_\mathrm{d}$, modeled by an electric dipole $\mathbf{p}_\mathrm{d}$, is embedded in medium (ε_2, μ_2) and illuminated by a plane EM wave of frequency ω

case that of a spherical NP, allows us to calculate the decay and excitation rate of the QE, as well as a number of relevant quantities that we discuss in the following paragraphs.

Excitation Enhancement In the long-wavelength limit, the interplay between a QE and light is described by the interaction Hamiltonian $\mathscr{H}_\mathrm{int} = -\mathbf{p}_\mathrm{d} \cdot \mathbf{E}$, i.e. a term of coupling of the electric dipole moment of the emitter with the external electric field. According to Fermi's golden rule, the excitation rate of the emitter is proportional to the square of the norm of the Hamiltonian element for the transition between the initial i and final state f, as [14]

Box 2.16 | Fermi's golden rule

$$\gamma_{i\rightarrow f} = \frac{2\pi}{\hbar^2}\left|\langle f|\mathscr{H}_\mathrm{int}|i\rangle\right|^2 \delta(\omega_i - \omega_f)\,. \tag{2.107}$$

In free space, the electric field responsible for the excitation is just the incident light $\mathbf{E}_0$. In the presence of a scatterer, the total field $\mathbf{E}_\mathrm{tot}$ contains additionally a scattered-field component in the space surrounding the NP. The excitation-rate enhancement, below saturation,[24] is then equal to

$$\eta_\mathrm{exc} = \frac{\gamma_\mathrm{exc}}{\gamma^0_\mathrm{exc}} = \frac{|\mathbf{p}_\mathrm{d} \cdot \mathbf{E}_\mathrm{tot}(\mathbf{r}_\mathrm{d})|^2}{|\mathbf{p}_\mathrm{d} \cdot \mathbf{E}_0(\mathbf{r}_\mathrm{d})|^2}\,. \tag{2.108}$$

Photonic Enhancement The infinitesimal work exerted on a charge q traveling a distance $d\mathbf{l} = \mathbf{v}dt$ in the presence of an EM field is given by

$$dW = \mathbf{F} \cdot d\mathbf{l} = q(\mathbf{E} + \mathbf{v} \times \mathbf{B}) \cdot \mathbf{v}dt = q\mathbf{E} \cdot \mathbf{v}dt = \mathbf{J} \cdot \mathbf{E}\,dt\,. \tag{2.109}$$

[24] At high light intensities, the excitation rate becomes larger than the spontaneous emission rate. However, there is a finite number of electrons that can be excited, beyond which any additional pumping does not prompt any changes to the system.

We can easily calculate the total power dissipated by the oscillating dipole over an oscillating period as

$$P = \left\langle \frac{dW}{dt} \right\rangle = -\frac{1}{2} \int dt \, \mathrm{Re}\{\mathbf{J}^*(\mathbf{r}) \cdot \mathbf{E}(\mathbf{r})\} = \frac{\omega^3}{2c^2\varepsilon_2\varepsilon_0} \mathrm{Im}\{\mathbf{p}_\mathrm{d}^* \cdot \overleftrightarrow{\mathscr{G}}(\mathbf{r}, \mathbf{r}_\mathrm{d}, \omega) \cdot \mathbf{p}_\mathrm{d}\} \, . \tag{2.110}$$

The ratio P/P_0 of the total dissipated power over the dissipated power in free space [27]

$$P_0 = \frac{ck_2^4 p_\mathrm{d}^2}{12\pi\varepsilon_2\varepsilon_0\sqrt{\varepsilon_2\mu_2}} \tag{2.111}$$

expresses the photonic enhancement

$$\frac{P}{P_0} = 1 + \frac{6\pi}{k_2} \mathrm{Im}\{\mathbf{p}_\mathrm{d}^* \cdot \overleftrightarrow{\mathscr{G}}_\mathrm{ref}(\hat{\mathbf{r}}, \mathbf{r}_\mathrm{d}, \omega) \cdot \hat{\mathbf{p}}_\mathrm{d}\} \tag{2.112}$$

due to the presence of the NP. Despite its name, this ratio can assume values smaller than one, indicating suppression of the dissipated power. Only part of the dissipated power is emitted in the form of radiation, while the remaining part is absorbed in the photonic environment, and is eventually converted to heat. We can thus divide the total dissipation power into radiative and non-radiative losses, $P = P_\mathrm{r} + P_\mathrm{nr}$.

Decay Rate Assuming that the QE emits photons of energy $\hbar\omega$, the radiative and non-radiative decay rates are defined as the ratio of the corresponding power over the photon energy

$$\gamma_{\mathrm{r/nr}} = \frac{P_{\mathrm{r/nr}}}{\hbar\omega} \, . \tag{2.113}$$

In free space, the decay of the QE is the result of not only radiation but also of intrinsic losses (many-photon processes, electron-transfer processes, etc.), and therefore the total decay rate can be expressed as the sum of the decay rates due to both radiative and non-radiative mechanisms, $\gamma_0 = \gamma_{0\mathrm{r}} + \gamma_{0\mathrm{nr}}$. The presence of the scatterer modifies the radiative decay rate $\gamma_{0\mathrm{r}}$ (the internal relaxation $\gamma_{0\mathrm{nr}}$ usually remains unaffected) while also introducing an additional non-radiative decay mechanism, i.e. energy absorption. Then the total decay rate is equal to

$$\gamma = \gamma_\mathrm{r} + \gamma_\mathrm{nr} + \gamma_{0\mathrm{nr}} = \frac{P}{\hbar\omega} \, , \tag{2.114}$$

where P is the total dissipated power given in Eq. (2.110).

Quantum Yield The quantum yield of the QE is a measure of its efficiency of converting absorbed light into emitted light. In free space, the characteristic internal quantum yield of the emitter is equal to $\mathscr{Q}_0 = \gamma_{0\mathrm{r}}/(\gamma_{0\mathrm{r}} + \gamma_{0\mathrm{nr}})$. In the presence of the scatterer, it is modified to $\mathscr{Q} = \gamma_\mathrm{r}/(\gamma_\mathrm{r} + \gamma_\mathrm{nr} + \gamma_{0\mathrm{nr}})$, or

$$\mathcal{Q} = \frac{\gamma_r/\gamma_{0r}}{\gamma_r/\gamma_{0r} + \gamma_{nr}/\gamma_{0r} + (\mathcal{Q}_0^{-1} - 1)} = \frac{P_r/P_{0r}}{P_r/P_{0r} + P_{nr}/P_{0r} + (\mathcal{Q}_0^{-1} - 1)} , \tag{2.115}$$

where P_{0r} is the radiated power of the emitter in free space [35].

Fluorescence Enhancement The number of photons emitted per unit time is described by the fluorescence rate $\gamma_{em} = \mathcal{Q}\eta_{exc}$. When studying fluorescence enhancement (or suppression), we can assume that the excitation and emission processes do not affect each other, and thus treat them separately. The total fluorescence enhancement is thus [36]

$$\eta_{em} = \frac{\gamma_{em}}{\gamma_{0em}} = \eta_{exc}\frac{\mathcal{Q}}{\mathcal{Q}_0} . \tag{2.116}$$

2.3.5 Mesoscopic Plasmonics

Our discussion so far has been limited to the classical local theory of electrodynamics, formulated by three assumptions: the macroscopic Maxwell equations, the linear and local constitutive relations, and the "hard-wall" boundary conditions. While, in most settings, this framework would be enough to predict the electric and optical properties of matter, it inevitably fails in predicting the physics governing nanometric metal objects. To begin with, contrary to the locality assumption, perturbations have a finite range, thus no system responds truly locally. Revisiting our model, the nonlocal response of a linear isotropic medium is reflected in its dielectric function connecting the response $\mathbf{D}(\mathbf{r})$ of the nonlocal material at point $\mathbf{r}$ to the applied electric field $\mathbf{E}(\mathbf{r}')$ at any other point $\mathbf{r}'$ through the constitutive relations (2.8), as

$$\mathbf{D}(\mathbf{r}, \omega) = \varepsilon_0 \int d\mathbf{r}' \, \varepsilon(\mathbf{r}, \mathbf{r}', \omega)\mathbf{E}(\mathbf{r}', \omega) . \tag{2.117}$$

Once the length scales involved become commensurate to a characteristic intrinsic length of the system, the local dielectric functions prove inadequate for the full description of the material. In metals this length is directly related to the Fermi wavelength[25] or the electron mean free path [8]. When the dimensions of the structure exceed these critical values, all effects pertaining to nonlocality are masked by the bulk response. At the same time, surface effects of quantum mechanical origin are overlooked due to the hard-wall boundary assumption. In practice, metals have finite work functions, meaning that there is a non-zero probability of electron spilling out of the surface of the metal. As a consequence, the electron density does not follow a strictly step-like profile as the one shown in Fig. 2.9a (red dashed line). On the contrary, it exhibits an evanescent tail, as well as Friedel oscillations, resulting from the interference of standing waves formed due to the termination of the metal [8].

[25] The Fermi wavelength is the de Broglie wavelength $\lambda_F = 2\pi/k_F$ of the electrons present near the Fermi energy level [8].

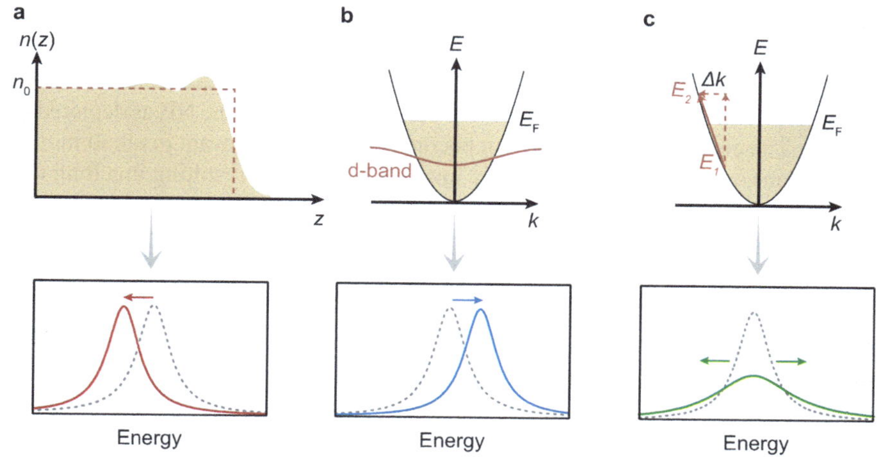

Fig. 2.9 Nonlocal and finite-size effects. a The equilibrium electron density n (shaded area) of the metal spills out of the assumed boundary (dashed line), leading to a redshift of the plasmon mode. **b** Band structure of a metal showing the d-band lying lower than the Fermi level resulting in screening. This mechanism is responsible for the blueshift of the plasmon modes observed in noble metals. **c** Landau damping describes the mechanism where an electron of energy $E_1 < E_F$, upon absorbing a photon, gains momentum $\Delta\mathbf{k}$ (from collision with the metal boundary, in the case of *surface-enabled* Landau damping), and is promoted to an energy state $E_2 > E_F$, thereby creating an electron-hole pair. This additional loss mechanism manifests as a linewidth-broadening of the plasmon resonance

Table 2.1 Comparison of nonlocal models

	SDB	HDM	GNOR	SRF
Permittivity	Local	Nonlocal	Nonlocal	Local
Boundary	Hard-wall	Hard-wall	Hard-wall	Transition domain
Corrections	Size-dependent broadening	Convection	Convection and diffusion	Induced surface polarization
Screening	✗	✓	✓	✓
Spill-out	✗	✗	✗	✓
Landau damping	✓	✗	✓	✓

In the following sections, we review some of the effects emerging in nanoscale systems and expose the inefficiency of LRA, building upon the discussion published in Publication C. We then discuss four approaches developed in the field of the so-called *mesoscopic plasmonics* in an effort to remedy these shortcomings [37]; the Kreibig size-dependent broadening (SDB) model, HDM, the generalized nonlocal optical response model (GNOR) and the Feibelman surface-response formalism (SRF). Table 2.1 provides a brief overview of the main assumptions underlying each model, as well as their respective strengths and limitations.

2.3.5.1 Nonlocal and Finite-Size Effects

Spill Out Due to their finite work functions, conduction electrons have a non-negligible probability of surmounting the assumed boundary of the NP, as depicted in Fig. 2.9a. The electron density spill-out becomes especially relevant in alkali metals (e.g. sodium) owing to their particularly low work functions, resulting in a long (on the order of a few Å) low-electron-density tail lying beyond the classical boundary. This manifests as a spectral redshift of the plasmonic modes, since the NP behaves as effectively larger. In ultranarrow gaps, e.g. between two nearly touching metal NPs, spill-out can even lead to direct electron tunneling from one NP to the other [38].

Screening The spatial dispersion of the dielectric function associated with nonlocality translates into wavevector dependence [39], which accounts for the effect of screening: the induced charges have some spatial extent, and inner electrons screen the interaction with the outer ones, leading to the NP behaving as effectively smaller. It should be noted that the NPs under study are so small that they can be considered as lying in the quasistatic regime, and frequency shifts should not be observed; nevertheless, the moment quantum effects related to the extreme nanoscale are considered, the quasistatic approximation is no longer valid, as we are effectively examining the intrinsic properties of the assumed point dipole. Screening is a competing mechanism to electron spill-out. In noble metals (e.g. silver and gold) that feature high work functions and large contributions from d-band screening (see Fig. 2.9b), this is the dominant mechanism, resulting in the electron density spilling inside the assumed boundary, and, consequently, in the blueshift of the plasmon resonances.

Landau Damping Figure 2.9c provides a simplistic picture of the band structure in a metal and the origin of this damping mechanism. Assume two energy states, one of energy E_1 below the Fermi level, and a state $E_2 = \hbar\omega + E_1$ above it, characterized by wavevectors $\mathbf{k}_1$ and $\mathbf{k}_2 = \mathbf{k}_1 + \Delta\mathbf{k}$, accordingly. Typically, an incident photon carrying a wavenumber $k \approx 0$, or an SPP, would not suffice for the direct transition from state 1 to state 2 [40]. But collisions with the boundaries of the NP can provide the required momentum $\Delta k \geq \omega/v_{\mathrm{F}}$ for the photon absorption, where v_{F} is the Fermi velocity.[26] As a consequence, electrons are promoted to the higher energy state, and the plasmon decays into electron–hole pairs. This additional energy dissipation channel appears as a broadening of the plasmon resonance in the spectra.

Since Landau damping is associated with electron–boundary collisions, it goes without saying that it most likely occurs near the surface of the metal, rather than in the bulk, justifying, thus, the full term *surface-enhanced/enabled Landau damping*. Naturally, it becomes increasingly more apparent as the NP size reduces, coexisting with—and maybe dominating over—other relaxation mechanisms. References [40, 41] provide an interesting discussion regarding the role of Landau damping in small metallic structures as the physical limit of plasmonic field confinement and enhancement.

---

[26] The Fermi velocity is the speed of the fastest electrons that occupy the Fermi level at absolute zero temperature.

2.3.5.2 The Kreibig Model

The increased damping in small metallic spheres was observed quite a few decades ago and studied more extensively by Kreibig and colleagues, who ascribed it to a decreased electron mean free path due to scattering off the NP surface [42], describing essentially the same process as Landau damping. Kreibig's postulation boils down to the addition of a phenomenological correction to the damping parameter [43], that describes the rate at which electrons collide to the metal walls,

$$\gamma \rightarrow \gamma_{\mathrm{bulk}} + A\,\frac{v_{\mathrm{F}}}{R}\,, \tag{2.118}$$

where A is a dimensionless constant (empirically $A \approx 1$), and R the radius of the spherical NP, or any other characteristic dimension of a non-spherical configuration. In practice, the inverse-proportionality relation $\gamma \propto 1/R$ reflects the surface-to-volume ratio S/V; in the same vein as Landau damping, in the bulk limit the surface collision damping rate is generally negligible, but it is enhanced near the surface. In sufficiently small NPs, where the mean free path of the electrons is comparable to the dimension of the structure $R \sim v_F/\gamma_{\mathrm{bulk}}$, this term becomes comparable to the other bulk contributions.

2.3.5.3 The Hydrodynamic Drude Model

As already briefly discussed in Sect. 2.2.2.3, HDM, proposed back in the 1930s [11], treats the dynamics of the metal as that of a compressible electron gas, and encompasses nonlocal corrections while maintaining a CED core. Consider a metallic medium with optical parameters ε_{m} and $\mu_{\mathrm{m}} = 1$. Reworking the equations of motion (2.64), while also introducing the current density $\mathbf{J} = en_0\mathbf{v}$, we arrive at the generalized Ohm's law for HDM,

$$\frac{\beta^2}{\omega\,(\omega + i\gamma)}\nabla\left[\nabla\cdot\mathbf{J}(\mathbf{r},\omega)\right] + \mathbf{J}(\mathbf{r},\omega) = \sigma(\omega)\mathbf{E}(\mathbf{r},\omega)\,, \tag{2.119}$$

where σ is the Drude conductivity of Eq. (2.51). The first term in Eq. (2.119) embodies the nonlocal correction to Ohm's law, that naturally vanishes as $\beta \to 0$. For a three-dimensional electron gas, the hydrodynamic parameter takes the value $\beta^2 = v_{\mathrm{F}}^2/3$ in the low-frequency limit ($\omega \ll \gamma$), and $\beta^2 = 3v_{\mathrm{F}}^2/5$ in the high-frequency limit ($\omega \gg \gamma$) which is relevant to plasmonics [44],[27] as one can derive through the total energy of the FEG [9]. If we use Eq. (2.19), we can eliminate $\mathbf{J}$ in Eq. (2.119) and obtain a generalized nonlocal wave equation

[27] More information regarding the value of the hydrodynamic parameter β in arbitrary frequencies can be found in Ref. [44].

Box 2.17 | Generalized vector wave equation for HDM

$$\nabla \times \nabla \times \mathbf{E} - k_0^2 \varepsilon_m \mathbf{E} = k_0^2 \xi_{NL} \nabla (\nabla \cdot \mathbf{E}) \ , \tag{2.120}$$

where ξ_{NL} is the nonlocal length defined as

$$\xi_{NL}^2 = \frac{\varepsilon_\infty}{\omega(\omega + i\gamma)} \beta^2 \ . \tag{2.121}$$

The above equation reproduces the transverse- and longitudinal-wave solutions that we derived through the connection of the polarization to the electric field in Sect. 2.1.6; the former corresponds to plane waves of wavevector $k_{m\perp} = k_0\sqrt{\varepsilon_m}$, whereas the wavevector associated with the longitudinal electric field is given by $k_{m\parallel} = \sqrt{\varepsilon_m}/\xi_{NL}$, and it can be obtained from the dispersion law for longitudinal waves $\varepsilon_\parallel = 0$, combined with the dielectric function of Eq. (2.68b).

The presence of longitudinal waves in the electric field calls for an additional boundary condition. Assuming a step profile for the equilibrium electron density, i.e. n_0 constant inside the metal and abruptly dropping to zero at the metal boundary $\partial\Omega$, leads to the condition [45]

$$\mathbf{J} \cdot \hat{\mathbf{n}}\Big|_{\mathbf{r} \in \partial\Omega} = 0 \ , \tag{2.122}$$

with $\hat{\mathbf{n}}$ denoting the unit vector normal to the metal surface. This essentially "hard-wall" boundary assumption is by construction responsible for the inability of HDM to describe electron spill-out beyond the assumed surface of the metal (as shown schematically in Fig. 2.9). Extensions to HDM have been recently developed that relax this constraint. For instance, in the self-consistent HDM, corrections in regions of intense variation of the equilibrium electron density, such as close to the metal boundary, are taken into account and electron spill-out can be predicted successfully [46–48], albeit at a higher computational cost.

A first attempt to incorporate longitudinal waves in Mie theory was made by Ruppin [49], who calculated the HDM-modified Mie coefficients, as

$$T_{E\ell} = \frac{\varepsilon_m j_\ell(k_m R)\Psi'_\ell(k_d R) - \varepsilon_d \Psi'_\ell(k_m R) j_\ell(k_d R) - \delta_\ell^{NL} j_\ell(k_m R) j_\ell(k_d R)}{\varepsilon_d h_\ell^+(k_d R)\Psi'_\ell(k_m R) - \varepsilon_m \xi'_\ell(k_d R) j_\ell(k_m R) + \delta_\ell^{NL} j_\ell(k_m R) h_\ell^+(k_d R)} \ , \tag{2.123a}$$

$$T_{M\ell} = \frac{j_\ell(k_m R)\Psi'_\ell(k_d R) - \Psi'_\ell(k_m R) j_\ell(k_d R)}{h_\ell^+(k_d R)\Psi'_\ell(k_m R) - \xi'_\ell(k_d R) j_\ell(k_m R)} \ , \tag{2.123b}$$

where ε_d is the dielectric function of the medium surrounding the metal. Equations (2.123) differ from the equivalent expressions in LRA (2.77) only in the $T_{E\ell}$ coefficient, by a nonlocal correction

$$\delta_\ell^{NL} = \ell(\ell + 1)\,\varepsilon_d \left(\frac{\varepsilon_m}{\varepsilon_\infty} - 1\right) \frac{j_\ell(k_{m\parallel} R)}{k_{m\parallel} R j'_\ell(k_{m\parallel} R)} \ . \tag{2.124}$$

In the limit of small radii, the poles of Eq. (2.123a) yield an expression for the LSP frequencies, equivalent to Eq. (2.81)

$$\omega_\ell = \sqrt{\frac{\ell}{\ell\varepsilon_\infty + (\ell+1)\,\varepsilon_\mathrm{d} + \delta_\ell^\mathrm{NL}}}\,\omega_\mathrm{p}\,. \tag{2.125}$$

2.3.5.4 The Generalized Nonlocal Optical Response Model

Despite its merits, HDM still fails to describe crucial nonlocal phenomena, Landau damping being a characteristic example. To account for the latter, GNOR model, developed by Mortensen et al. [50], adds to the convective contribution of HDM a diffusion term, described by the diffusion constant $\mathscr{D}$, that acts as a relaxation mechanism, emulating surface-scattering phenomena. This term modifies the continuity equation in the frequency domain to

$$-i\omega e n(\mathbf{r},\omega) = \mathscr{D}\nabla^2\left[en(\mathbf{r},\omega)\right] + \boldsymbol{\nabla}\cdot\left[-en_0\mathbf{v}(\mathbf{r},\omega)\right] = \boldsymbol{\nabla}\cdot\mathbf{J}(\mathbf{r},\omega)\,. \tag{2.126}$$

The current density acquires an additional diffusion term, according to Fick's law[28]

$$\mathbf{J}(\mathbf{r},\omega) = -en_0\mathbf{v}(\mathbf{r},\omega) + e\mathscr{D}\boldsymbol{\nabla}n(\mathbf{r},\omega)\,, \tag{2.127}$$

which leads to the generalized Ohm's law within GNOR,

$$\left[\frac{\beta^2}{\omega\,(\omega+i\gamma)} + \frac{\mathscr{D}}{i\omega}\right]\nabla\left[\boldsymbol{\nabla}\cdot\mathbf{J}(\mathbf{r},\omega)\right] + \mathbf{J}(\mathbf{r},\omega) = \sigma(\omega)\mathbf{E}(\mathbf{r},\omega)\,. \tag{2.128}$$

This looks remarkably similar to Eq. (2.119), except for the $\mathscr{D}/i\omega$ term in the square brackets on the left-hand side. Most computational tools developed for HDM can thus be extended to GNOR, simply by introducing the complex hydrodynamic convective-diffusive parameter $\eta^2 = \beta^2 + \mathscr{D}\,(\gamma - i\omega)$ in place of β^2.

2.3.5.5 The Feibelman Surface-Response Formalism

A more efficient treatment of the hard-wall constraint is to invoke surface-response functions at the metal boundary. The step profile of the electron density, prior consisting of the metal and the host environment parts, now breaks down into an additional domain between the two: a transition region characterized by an evanescent electron-density tail and Friedel oscillations (see Fig. 2.10a). In the framework of Feibelman's

[28] Fick's law of diffusion postulates that the flux goes from regions of high concentration to regions of low concentration as $\mathbf{J} = -\mathscr{D}\boldsymbol{\nabla}n$ [51].

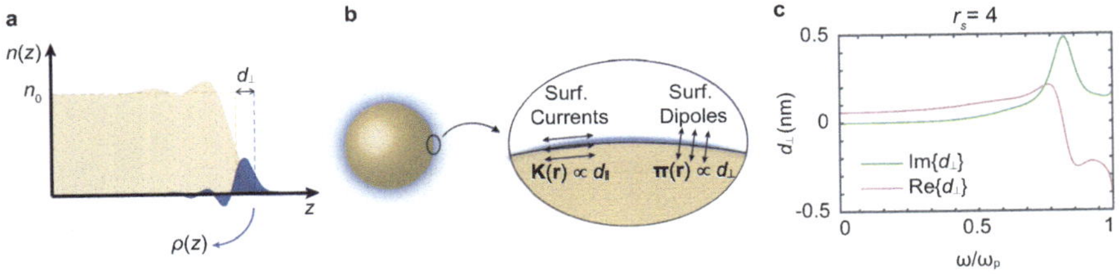

Fig. 2.10 The Feibelman SRF. a The electron-density profile containing a transition region characterized by an evanescent electron-density tail and Friedel oscillations (yellow area). The $d_\perp$ parameter corresponds to the position of the centroid of the induced surface charge density (blue area) with respect to the jellium boundary of the structure (yellow dashed line). **b** Schematic illustration of the surface induced charges and currents associated with the Feibelman d-parameters. **c** The real and imaginary part of $d_\perp$ as a function of frequency normalized to ω_p for a jellium metal with Wigner-Seitz radius $r_s = 4$ (in units of the Bohr radius), that can describe sodium well [57]

work in surface science [52], nonlocality is imposed via the surface-response parameters $d_\perp$ and $d_\parallel$, also called Feibelman d-parameters, entering the ordinary boundary conditions, while the bulk material is still described by local response functions.[29] In what follows, we only provide a brief overview of the Feibelman SRF, as detailed derivations can be found elsewhere [54, 55].

As illustrated in Fig. 2.10b, $d_\perp$ and $d_\parallel$ represent the frequency-dependent centroids of the induced charge ρ and the normal derivative of the tangential component J_x of the current density, respectively, evaluated at a planar metal–dielectric interface (assumed to lie in the xy plane). In mathematical notation this translates to

Box 2.18 | Feibelman d-parameters

$$d_\perp(\omega) = \frac{\int dz\, z\rho(z,\omega)}{\int dz\, \rho(z,\omega)}, \qquad d_\parallel(\omega) = \frac{\int dz\, z\partial_z J_x(z,\omega)}{\int dz\, \partial_z J_x(z,\omega)}, \tag{2.129}$$

with $d_j(\omega) = d'_j(\omega) + i d''_j(\omega)$ ($j = \perp, \parallel$) being in general complex-valued. For $d_\perp$, the real part describes the effective position of the centroid of induced surface charge with respect to the assumed interface [56]. Positive $d'_\perp$ means spill-out of the density profile, while negative values correspond to spill-in (see Fig. 2.10a), although this can depend on the initial assumption of where the interface is located. The diffuseness of the density profile, at the same time, influences the probability of generating electron-hole pairs in the surface region due to Landau damping. This property is reflected in the imaginary part $d''_\perp$.

---

[29] There are certainly more than one routes to treat the transition domain, for instance as an infinitely thin layer of dipoles oriented normally to the NP surface [53].

The Feibelman parameters of Eq. (2.129) can be incorporated in the boundary conditions through a (normally-oriented) surface polarization and a (tangentially-oriented) current term, $\mathbf{P}(\mathbf{r}) = \pi(\mathbf{r})\delta(\mathbf{r}-\mathbf{s})$ and $\mathbf{J}(\mathbf{r}) = \mathbf{K}(\mathbf{r})\delta(\mathbf{r}-\mathbf{s})$ accordingly, where $\mathbf{s}$ denotes points on the metal–dielectric interface. These terms are each proportional to a Feibelman d-parameter and arise from the discontinuities of the field components [54],

$$\pi = \varepsilon_0 d_\perp \left[\hat{\mathbf{n}} \cdot (\mathbf{E}_\mathrm{d} - \mathbf{E}_\mathrm{m})\right]\hat{\mathbf{n}}\,, \tag{2.130a}$$

$$\mathbf{K} = i\omega d_\parallel \left[\hat{\mathbf{n}} \times (\mathbf{D}_\mathrm{d} - \mathbf{D}_\mathrm{m}) \times \hat{\mathbf{n}}\right], \tag{2.130b}$$

where the subscript d refers to the corresponding field in the dielectric environment, and m to that in the metal. The densities can be absorbed into the boundary conditions of Eqs. (2.15), leading to [55]

$$\mathbf{E}_{\mathrm{d}\parallel} - \mathbf{E}_{\mathrm{m}\parallel} = -d_\perp \nabla_\parallel \left(E_{\mathrm{d}\perp} - E_{\mathrm{m}\perp}\right), \tag{2.131a}$$

$$\mathbf{H}_{\mathrm{d}\parallel} - \mathbf{H}_{\mathrm{m}\parallel} = i\omega d_\parallel \left(\mathbf{D}_{\mathrm{d}\parallel} - \mathbf{D}_{\mathrm{m}\parallel}\right) \times \hat{\mathbf{n}}\,, \tag{2.131b}$$

where $\nabla_\parallel$ is the surface nabla operator. Now, we can easily generalize Mie theory imposing the modified boundary conditions to find the Mie coefficients, which read

$$T_{\mathrm{E}\ell} = \frac{\varepsilon_\mathrm{m} j_\ell(k_\mathrm{m}R)\Psi'_\ell(k_\mathrm{d}R) - \varepsilon_\mathrm{d}\Psi'_\ell(k_\mathrm{m}R) j_\ell(k_\mathrm{d}R) + \Delta_\ell^\mathrm{num}}{\varepsilon_\mathrm{d} h_\ell^+(k_\mathrm{d}R)\Psi'_\ell(k_\mathrm{m}R) - \varepsilon_\mathrm{m}\xi'_\ell(k_\mathrm{d}R) j_\ell(k_\mathrm{m}R) - \Delta_\ell^\mathrm{den}}\,, \tag{2.132a}$$

$$T_{\mathrm{M}\ell} = \frac{j_\ell(k_\mathrm{m}R)\Psi'_\ell(k_\mathrm{d}R) - \Psi'_\ell(k_\mathrm{m}R) j_\ell(k_\mathrm{d}R) + R^2\left(k_\mathrm{m}^2 - k_\mathrm{d}^2\right) j_\ell(k_\mathrm{m}R) j_\ell(k_\mathrm{d}R)\widetilde{d}_\parallel}{h_\ell^+(k_\mathrm{d}R)\Psi'_\ell(k_\mathrm{m}R) - \xi'_\ell(k_\mathrm{d}R) j_\ell(k_\mathrm{m}R) - R^2\left(k_\mathrm{m}^2 - k_\mathrm{d}^2\right) j_\ell(k_\mathrm{m}R) h_\ell(k_\mathrm{d}R)\widetilde{d}_\parallel}\,. \tag{2.132b}$$

In Eqs. (2.132) we have set the dimensionless parameters $\widetilde{d}_\perp = \ell(\ell+1)d_\perp/R$ and $\widetilde{d}_\parallel = d_\parallel/R$, as well as

$$\Delta_\ell^\mathrm{num} = (\varepsilon_\mathrm{m} - \varepsilon_\mathrm{d})\left[j_\ell(k_\mathrm{m}R) j_\ell(k_\mathrm{d}R)\widetilde{d}_\perp + \Psi'_\ell(k_\mathrm{m}R)\Psi'_\ell(k_\mathrm{d}R)\widetilde{d}_\parallel\right], \tag{2.133a}$$

$$\Delta_\ell^\mathrm{den} = (\varepsilon_\mathrm{m} - \varepsilon_\mathrm{d})\left[j_\ell(k_\mathrm{m}R) h_\ell^+(k_\mathrm{d}R)\widetilde{d}_\perp + \Psi'_\ell(k_\mathrm{m}R)\xi'_\ell(k_\mathrm{d}R)\widetilde{d}_\parallel\right]. \tag{2.133b}$$

To compare with the previous models, SRF predicts that the frequencies of the LSPs of a small sphere are given, in analogy with Eq. (2.80), by [55]

$$\varepsilon_\mathrm{m} + \frac{\ell+1}{\ell}\varepsilon_\mathrm{d} - (\varepsilon_\mathrm{m} - \varepsilon_\mathrm{d})\frac{\ell+1}{R}\left(d_\perp - d_\parallel\right) = 0\,. \tag{2.134}$$

Unlike Eq. (2.80), this expression cannot be straightforwardly solved for the LSP frequencies ω_ℓ, since the d-parameters are also frequency-dependent. One can solve it numerically or graphically for given surface-response functions, or derive a perturbative solution for small shifts from the LRA result. The latter approach, assuming $\varepsilon_\mathrm{d} = \varepsilon_\infty = 1$, results in [55]

$$\omega_\ell \simeq \omega_{\ell,\mathrm{LRA}} \left[1 - \frac{1}{2} \left(\frac{\ell+1}{R} d_\perp^{(0)} - \frac{\ell+1}{R} d_\parallel^{(0)} \right) \right] , \tag{2.135}$$

where $\omega_{\ell,\mathrm{LRA}}$ are the frequencies given by Eq. (2.81), and $d_{\perp,\parallel}^{(0)}$ corresponds to the real part of the corresponding function calculated at $\omega_{\ell,\mathrm{LRA}}$. For the evaluation of the d-parameters several methods have been proposed, with the most popular being time-dependent density functional theory (TDDFT) calculations for a flat metal-air interface. In this approach, the metal is treated as a *jellium* medium (or homogeneous electron gas), meaning that the electrons experience a uniformly distributed ionic background, and only interact with each other. Figure 2.10c, shows $d_\perp$ as a function of frequency for a jellium metal for a Wigner-Seitz radius $r_s = 4$ in units of the Bohr radius, which is well-suited, e.g., for sodium.[30] Other approaches include a smoothly varying electron density function within LRA [58], or an atomic-layer potential as the starting point [59].

As a final comment on the Feibelman approach, it should be emphasized that it is an approximation based on a number of assumptions, and its validity is constrained to the long-wavelength approximation. Firstly, the d-parameters emerge from a multipole expansion of the quantum mechanical charge and current density, where each multipole term n contributes $(k_\parallel x)^n$, with $k_\parallel$ denoting the in-plane wavevector at a planar interface and x is a length scale. The monopole term reproduces LRA, the dipole term yields $x = d_{\perp/\parallel}$ [Eqs. (2.129)], while the higher order terms are neglected. For a planar interface, this truncation holds only as long as $1/k_\parallel \gg |d_{\perp/\parallel}|$, i.e. under the assumption that the field variation along the metallic surface is much larger than the correction. The equivalent condition for a mode of order ℓ confined at a spherical surface of radius R is $R/\ell \gg |d_{\perp/\parallel}| \sim$ Å.[31] This condition also assures that the d-parameters derived for a planar metal-air interface can be used at curved surfaces, since they experience it as locally flat. Finally, within the long-wavelength approximation we neglect the nonlocality of the d-parameters [60].

2.4 Numerical Methods

Large part of this thesis is devoted to spherical geometries that allow to study most EM problems with closed-form analytic expressions based on Mie-type calculations. Undeniably, analytics provide the most accurate results at low computational cost, while also offering a higher level of controlability and deeper understanding of the underlying physics. However, exact solutions can only be derived for a handful of simple planar, spherical, and cylindrical geometries, that most often do not emulate accurately experimental setups.

---

30 We use the values provided in Ref. [55], which were generated by means of a multi-Lorentzian fitting of the TTDFT data derived in Ref. [57].

31 At a planar interface the surface charge has periodicity length $2\pi/k_\parallel$, whereas for a spherical one it is equivalent to $2\pi R/\sqrt{\ell(\ell+1)}$.

To complement the shortcomings of analytical methods, several numerical techniques have been developed for solving Maxwell's equations. Numerical simulations offer versatility and better modeling of experimental practices. Here we discuss the discontinuous Galerkin time-domain method (DGTD). This technique is based on the concept of decomposing the spatial domain containing the structure under study into a finite number of smaller discrete elements, and utilizes the Galerkin scheme for solving the time-dependent Maxwell equations at each of these elements.

2.4.1 The Galerkin Scheme

The Galerkin scheme is employed to convert a continuous operator equation, e.g. a differential equation, to a discrete problem in a matrix representation that can be solved by matrix algebra. Assume that we are seeking a solution $v(\mathbf{r})$ within a three-dimensional domain $\mathcal{D}$ that satisfies the integral equation

$$\int_{\mathcal{D}} d^3r' K(\mathbf{r}, \mathbf{r}')v(\mathbf{r}') = b(\mathbf{r}) \,, \tag{2.136}$$

for some inhomogeneity b, where K is the kernel of the equation. Our goal is to find an approximate solution u to the exact solution v. We first expand the approximate solution to a truncated basis Φ_ν of dimension n

$$v(\mathbf{r}) \approx u(\mathbf{r}) = \sum_{\nu=1}^{n} u_\nu \Phi_\nu(\mathbf{r}) \,, \tag{2.137}$$

where u_ν are the expansion coefficients. Due to the truncated basis, Eq. (2.136) does not hold exactly for u, so once inserted back to the original equation, the approximate solution leads to a finite residuum

$$\int_{\mathcal{D}} d^3r' \left[K(\mathbf{r}, \mathbf{r}') \sum_{\nu=1}^{n} u_\nu \Phi_\nu(\mathbf{r}') \right] - b(\mathbf{r}) = \text{res} \,. \tag{2.138}$$

We now seek for the optimal solution u within the truncated function space. To this end, we multiply Eq. (2.138) with a test function

$$W(\mathbf{r}) = \sum_{\nu=1}^{n} w_\nu \Phi_\nu(\mathbf{r}) \,, \tag{2.139}$$

and we obtain

$$\sum_{\nu=1}^{n} w_\nu \left[\sum_{\nu'=1}^{n} K_{\nu\nu'} u_{\nu'} - b_\nu \right] = \sum_{\nu=1}^{n} w_\nu \int_{\mathcal{D}} d^3r \, \Phi_\nu(\mathbf{r}) \, \text{res} \,, \tag{2.140}$$

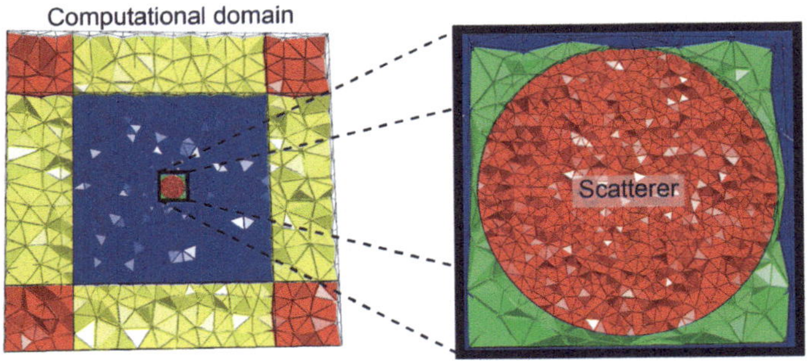

Fig. 2.11 Computational domain in the DGTD method. The volume of the structure and its surroundings are discretized in finite elements that make up the mesh. The fields are approximated individually at each element, and the solutions are subsequently stitched together in neighboring elements via a numerical flux scheme. Figure courtesy of Christian Wolff

where we have set

$$K_{\nu\nu'} = \int\int_{\mathcal{D}} d^3r d^3r' \Phi_\nu(\mathbf{r}) K(\mathbf{r}, \mathbf{r}') \Phi_{\nu'}(\mathbf{r}') \,, \tag{2.141a}$$

$$b_\nu = \int_{\mathcal{D}} d^3r \Phi_\nu(\mathbf{r}) b(\mathbf{r}) \,. \tag{2.141b}$$

The optimal solution is the one that minimizes the residuum, i.e. when the term in brackets in Eq. (2.140) is zero, giving $\mathbf{K} \cdot \mathbf{u} = \mathbf{b}$. Consequently, solving Eq. (2.136) reduces to a much simpler matrix inversion.

2.4.2 The Discontinuous Galerkin Time-Domain Method

The discontinuous Galerkin scheme is a variant of the finite element method (FEM). In this respect, it is a volume method, meaning that the field is evaluated all over the volume of the domain defined by the structure and its surroundings, as illustrated in Fig. 2.11. In order for the computational domain to remain finite, suitable boundary conditions need be applied on its boundary to mimic the physical behavior of the part of the system which cannot be simulated.

In the discontinuous Galerkin method, the EM field is expanded on a set of basis functions at each element τ_i that make up the mesh. In contrast to conventional FEM, where each element is coupled to its neighbors, in this method the basis functions are defined only locally and do not communicate to each other, hence the term 'discontinuous'. In other words,

$$\Phi_i(\mathbf{r}) = \begin{cases} \mathscr{L}_i(\mathbf{r}) , & \text{for } \mathbf{r} \in \mathcal{T}_i \\ 0 , & \text{elsewhere} , \end{cases} \tag{2.142}$$

where $\mathscr{L}_i$ represents a scalar basis function. The individual solutions are stitched together post hoc, under the requirement that neighboring elements must have the same flux, as we explain later on.

The DGTD method introduces a temporal discretization that enables us to solve electrodynamic problems, and generate spectra in the frequency domain via a Fourier transform. The time-domain analysis performed in this thesis is based on the DGTD method developed by Busch and coworkers [61], based on the work of Hasthaven and Warburton [62]. Here we present basic concepts of the algorithm and refer the reader to the aforementioned reference for technical details.

Let us start with the spatial discretization. Consider an EM field that satisfies Maxwell's equations written in their curl form

$$\varepsilon_0\varepsilon(\mathbf{r})\frac{\partial \mathbf{E}(\mathbf{r},t)}{\partial t} = \nabla \times \mathbf{H}(\mathbf{r},t) \tag{2.143a}$$

$$\mu_0\mu(\mathbf{r})\frac{\partial \mathbf{H}(\mathbf{r},t)}{\partial t} = -\nabla \times \mathbf{E}(\mathbf{r},t) , \tag{2.143b}$$

where we have assumed dispersionless and isotropic optical parameters. We may write Maxwell's equations as a conservation law

$$\mathscr{Z}(\mathbf{r}) \cdot \frac{\partial \mathbf{q}}{\partial t} + \nabla \cdot \mathbf{F}(\mathbf{q}) = 0 , \tag{2.144}$$

where we have introduced a state vector $\mathbf{q}(\mathbf{r},t) = [\mathbf{E}(\mathbf{r},t)\ \mathbf{H}(\mathbf{r},t)]^{\mathrm{T}}$, a flux vector , $\mathbf{F} = [\mathbf{F}_x\ \mathbf{F}_y\ \mathbf{F}_z]$, that depends on the tangential field components only

$$\mathbf{F}_i = \begin{pmatrix} -\hat{\mathbf{e}}_i \times \mathbf{H}(\mathbf{r},t) \\ \hat{\mathbf{e}}_i \times \mathbf{E}(\mathbf{r},t) \end{pmatrix} , \qquad \hat{\mathbf{e}}_i = \hat{\mathbf{x}}, \hat{\mathbf{y}}, \hat{\mathbf{z}} \tag{2.145}$$

and a material matrix

$$\mathscr{Z}(\mathbf{r}) = \begin{pmatrix} \varepsilon_0\varepsilon(\mathbf{r}) & 0 \\ 0 & \mu_0\mu(\mathbf{r}) \end{pmatrix} . \tag{2.146}$$

Consider a single element $\mathcal{T}_i$ of the computational domain, of volume V_i. The approximate solution $\mathbf{q}_\nu$ leaves a finite residuum when inserted back into Eq. (2.144). Within a given finite function space, there exists an optimal function that minimizes the residuum, which is, thus, orthogonal to this space. Then

$$\int_{V_i} d^3r\, \mathscr{L}_i(\mathbf{r})\Big[\mathscr{Z}(\mathbf{r}) \cdot \frac{\partial \mathbf{q}_\nu}{\partial t} + \nabla \cdot \mathbf{F}(\mathbf{q}_\nu)\Big] = \int_{V_i} d^3r\, \mathscr{L}_i(\mathbf{r}) \cdot \mathrm{res} = 0 , \tag{2.147}$$

for any test scalar basis function $\mathscr{L}_i$ of the function space. In the analyses included in this thesis the basis functions utilized are the Lagrange polynomials of order p. We stress that the above formula is local, restricted to holding for a single element only.

The coupling between neighboring elements is achieved by absorbing the continuity conditions at the boundary of each element into the physical equations. In the spirit of Ref. [61], we introduce a numerical flux $\widetilde{\mathbf{F}}$ to mediate the coupling, as

$$\int_{V_i} d^3r \mathscr{L}_i(\mathbf{r})\Big[\mathscr{Z}(\mathbf{r}) \cdot \frac{\partial \mathbf{q}_\nu}{\partial t} + \nabla \cdot \mathbf{F}(\mathbf{q}_\nu)\Big] = \int_{\partial V_i} da\, \mathscr{L}_i(\mathbf{r})\, \hat{\mathbf{n}} \cdot \big[\mathbf{F}(\mathbf{q}_\nu) - \widetilde{\mathbf{F}}(\mathbf{q}_\nu)\big] . \tag{2.148}$$

The consistence and convergence of the scheme depends on the numerical flux $\widetilde{\mathbf{F}}$, but its choice is not unique. Here we operate on the upwind flux scheme

$$\hat{\mathbf{n}} \cdot \big[\mathbf{F}(\mathbf{q}_\nu) - \widetilde{\mathbf{F}}(\mathbf{q}_\nu)\big] = \frac{\alpha}{\Delta Z}\big[\Delta \mathbf{E} - \hat{\mathbf{n}}(\hat{\mathbf{n}} \cdot \Delta \mathbf{E})\big] + \frac{Z^+}{\Delta Z}\hat{\mathbf{n}} \times \Delta \mathbf{H} , \tag{2.149}$$

where the superscript $+/-$ denotes the value of a quantity as we approach the boundary from outside/inside the corresponding element, Z is the impedance of the medium with $\Delta Z = Z^+ - Z^-$, $\Delta \mathbf{E} = \mathbf{E}^+ - \mathbf{E}^-$ is the discontinuity of the electric field at the boundary, and similarly for $\Delta \mathbf{H}$. This particular choice of flux has the advantage of strongly damping unphysical modes. It is clear, at this point, that in DGTD one has to store the fields at both sides of a boundary. The resulting matrices in DGTD, albeit larger compared to conventional FEM, they are block-diagonal, which simplifies the numerical calculation and reduces the computation time.

So far, we have employed the discontinuous Galerkin scheme to discretize the EM fields spatially. The temporal discretization is applied in terms of a time stepping scheme that evolves the fields from t_n to t_{n+1}. In DGTD the time evolution is achieved using the low-storage Runge-Kutta method [61]. Consider an equation of the general form

$$\frac{\partial \widetilde{\mathbf{q}}(t)}{\partial t} = f(\widetilde{\mathbf{q}}(t), t) . \tag{2.150}$$

Given a total number of ν expansion coefficients, the low-storage Runge-Kutta method requires a total of 2ν values stored in two registers $\widetilde{\mathbf{q}}$ and $\widetilde{\mathbf{p}}$. The algorithm from t_n to t_{n+1} operates as follows

$$\begin{aligned}
\widetilde{\mathbf{q}}_0 &:= \widetilde{\mathbf{q}}(t_n) \\
\widetilde{\mathbf{p}}_i &:= A_i \widetilde{\mathbf{p}}_{i-1} + \Delta t\, f(\widetilde{\mathbf{q}}_{i-1}, t_n + c_i \Delta t) . \\
\widetilde{\mathbf{q}}_i &:= \widetilde{\mathbf{q}}_{i-1} + B_i \widetilde{\mathbf{p}}_i \\
\widetilde{\mathbf{q}}(t_{n+1}) &:= \widetilde{\mathbf{q}}_s ,
\end{aligned}$$

where A_i, B_i and c_i define the properties of the scheme. The two middle steps define the stages $i = 1, \ldots, s$ of total number s.

References

1. Jackson JD (1998) Classical electrodynamics, 3rd ed. Wiley
2. Lindell I, Sihvola A, Tretyakov S, Viitanen AJ (1994) Electromagnetic waves in chiral and bi-isotropic media. Artech House
3. Barron LD (2009) Molecular light scattering and optical activity. Cambridge University Press
4. Hentschel M, Schäferling M, Weiss T, Liu N, Giessen H (2012) Nano Lett 12:2542. https://doi.org/10.1021/nl300769x
5. Lipkin DM (1964) J Math Phys 5:696. https://doi.org/10.1063/1.1704165
6. Tang Y, Cohen AE (2010) Phys Rev Lett 104:163901. https://doi.org/10.1103/PhysRevLett.104.163901
7. Mun J, Kim M, Yang Y, Badloe T, Ni J, Chen Y, Qiu C-W, Rho J (2020) Light Sci Appl 9:139. https://doi.org/10.1038/s41377-020-00367-8
8. Ashcroft NW, Mermin ND (1976) Solid state physics. Holt-Saunders
9. Griffiths DJ, Schroeter DF (2017) Introduction to quantum mechanics, 3rd ed. Cambridge University Press
10. Sommerfeld A (1927) Naturwissenschaften 15:825. https://doi.org/10.1007/BF01505083
11. Bloch F (1933) Z Phys 81:363. https://doi.org/10.1007/bf01344553
12. Raza S, Bozhevolnyi SI, Wubs M, Mortensen NA (2015) J Phys: Condens Matter 27:183204. https://doi.org/10.1088/0953-8984/27/18/183204
13. Yang W (1986) Phys Rev A 34:4575. https://doi.org/10.1103/PhysRevA.34.4575
14. Hohenester U (2020) Nano and quantum optics: an introduction to basic principles and theory. Springer
15. Stratton JA (1941) Electromagnetic theory. McGraw-Hill
16. Bohren CF, Huffman DR (1983) Absorption and scattering of light by small particles. Wiley
17. Ruppin R (2000) Opt Commun 182:273. https://doi.org/10.1016/S0030-4018(00)00825-7
18. Pines D, Bohm D (1952) Phys Rev 85:338. https://doi.org/10.1103/PhysRev.85.338
19. Ritchie RH (1957) Phys Rev 106:874. https://doi.org/10.1103/PhysRev.106.874
20. Powell CJ, Swan JB (1959) Phys Rev 115:869. https://doi.org/10.1103/PhysRev.115.869
21. Teng Y-Y, Stern EA (1967) Phys Rev Lett 19:511. https://doi.org/10.1103/PhysRevLett.19.511
22. Klimov V (2014) Nanoplasmonics. CRC Press
23. Green MA (2008) Sol Energy Mater Sol Cells 92:1305. https://doi.org/10.1016/j.solmat.2008.06.009
24. Tzarouchis DC, Ylä-Oijala P, Sihvola A (2016) Phys Rev B 94:140301(R). https://doi.org/10.1103/PhysRevB.94.140301
25. Kerker M, Wang D-S, Giles CL (1983) J Opt Soc Am 73:765. https://doi.org/10.1364/JOSA.73.000765
26. Fox M (2006) Quantum optics: an introduction. Oxford University Press
27. Novotny L, Hecht B (2006) Principles of nano-optics. Cambridge University Press
28. Mak KF, Lee C, Hone J, Shan J, Heinz TF (2010) Phys Rev Lett 105:136805. https://doi.org/10.1103/PhysRevLett.105.136805
29. Mueller T, Malic E (2018) Appl NPJ 2D Mater 2:29. https://doi.org/10.1038/s41699-018-0074-2
30. Holmes RJ, Forrest SR (2007) Org Electron 8:77. https://doi.org/10.1016/j.orgel.2006.05.005
31. Shi L, Hakala TK, Rekola HT, Martikainen J-P, Moerland RJ, Törmä P (2014) Phys Rev Lett 112:153002. https://doi.org/10.1103/PhysRevLett.112.153002
32. Törmä P, Barnes WL (2015) Rep Prog Phys 78:013901. https://doi.org/10.1088/0034-4885/78/1/013901
33. Otto A, Mrozek I, Grabhorn H, Akemann W (1992) J Phys: Condens Matter 4:1143. https://doi.org/10.1088/0953-8984/4/5/001
34. Purcell EM, Torrey HC, Pound RV (1946) Phys Rev 69:37. https://doi.org/10.1103/PhysRev.69.37

35. Bharadwaj P, Deutsch B, Novotny L (2009) Adv Opt Photon 1:438. https://doi.org/10.1364/AOP.1.000438
36. Anger P, Bharadwaj P, Novotny L (2006) Phys Rev Lett 96:113002. https://doi.org/10.1103/PhysRevLett.96.113002
37. Mortensen NA (2021) Nanophotonics 10:2563. https://doi.org/10.1515/nanoph-2021-0156
38. Savage KJ, Hawkeye MM, Esteban R, Borisov AG, Aizpurua J, Baumberg JJ (2012) Nature 491:574. https://doi.org/10.1038/nature11653
39. Lindhard J (1954) Dan Mat Fys Medd 28:1
40. Khurgin JB, Sun G (2016) Landau damping – the ultimate limit of field confinement and enhancement in plasmonic structures. In: Quantum plasmonics. Springer, pp 303–322
41. Khurgin J, Tsai W-Y, Tsai DP, Sun G (2017) ACS Photon 4:2871. https://doi.org/10.1021/acsphotonics.7b00860
42. Kreibig U, von Fragstein C (1969) Z Phys 224:307. https://doi.org/10.1007/BF01393059
43. Kreibig U, Vollmer M (1995) Optical properties of metal clusters. Springer
44. Halevi P (1995) Phys Rev B 51:7497. https://doi.org/10.1103/PhysRevB.51.7497
45. Melnyk AR, Harrison MJ (1970) Phys Rev B 2:835. https://doi.org/10.1103/PhysRevB.2.835
46. Toscano G, Straubel J, Kwiatkowski A, Rockstuhl C, Evers F, Xu H, Mortensen NA, Wubs M (2015) Nat Commun 6:7132. https://doi.org/10.1038/ncomms8132
47. Ciracì C, Della Sala F (2016) Phys Rev B 93:205405. https://doi.org/10.1103/PhysRevB.93.205405
48. Ciracì C (2017) Phys Rev B 95:245434. https://doi.org/10.1103/PhysRevB.95.245434
49. Ruppin R (1973) Phys Rev Lett 31:1434. https://doi.org/10.1103/PhysRevLett.31.1434
50. Mortensen NA, Raza S, Wubs M, Søndergaard T, Bozhevolnyi SI (2014) Nat Commun 5:3809. https://doi.org/10.1038/ncomms4809
51. Atkins P, de Paula J (2014) Atkins' physical chemistry, 10th ed. Oxford University Press
52. Feibelman PJ (1982) Prog Surf Sci 12:287. https://doi.org/10.1016/0079-6816(82)90001-6
53. Yan W, Wubs M, Mortensen NA (2015) Phys Rev Lett 115:137403. https://doi.org/10.1103/PhysRevLett.115.137403
54. Yang Y, Zhu D, Yan W, Agarwal A, Zheng M, Joannopoulos JD, Lalanne P, Christensen T, Berggren KK, Soljačić M (2019) Nature 576:248. https://doi.org/10.1038/s41586-019-1803-1
55. Gonçalves PAD, Christensen T, Rivera N, Jauho A-P, Mortensen NA, Soljačić M (2020) Nat Commun 11:366. https://doi.org/10.1038/s41467-019-13820-z
56. Liebsch A (1997) Electronic excitations at metal surfaces. Springer
57. Christensen T, Yan W, Jauho A-P, Soljačić M, Mortensen NA (2017) Phys Rev Lett 118:157402. https://doi.org/10.1103/PhysRevLett.118.157402
58. Mortensen NA, Gonçalves PAD, Shuklin FA, Cox JD, Tserkezis C, Ichikawa M, Wolff C (2021) Nanophotonics 10:3647. https://doi.org/10.1515/nanoph-2021-0084
59. Rodríguez Echarri Á, Gonçalves PAD, Tserkezis C, García de Abajo FJ, Mortensen NA, Cox JD (2021) Optica 8:710. https://doi.org/10.1364/OPTICA.412122
60. Babaze A, Neuman T, Esteban R, Aizpurua J, Borisov AG (2023) Nanophotonics 12:3277. https://doi.org/10.1515/nanoph-2023-0178
61. Busch K, König M, Niegemann J (2011) Laser Photon Rev 5:773. https://doi.org/10.1002/lpor.201000045
62. Hesthaven JS, Warburton T (2002) J Comput Phys 181:186. https://doi.org/10.1006/jcph.2002.7118

Chapter 3
Strong Coupling in Photonics

Abstract This chapter presents a study of strong coupling between optical excitations and electronic transitions sustained in excitonic materials, emerging in core—shell nanoparticles. Here we generalize Mie theory for isotropic core—shell structures and use it to analyze the coupling strength between metallic or dielectric nanocavities encapsulated in excitonic media. In the case of plasmon—exciton coupling, we examine the emerging modes as predicted by various nonclassical models within the field of mesoscopic plasmonics, and comment on the impact of nonclassical effects on the coupling strength. We, then, consider chiral nanocavities and investigate the optical properties of the resulting chiral polaritons.

In Sect. 2.3 we established that nanostructures combining optical transitions and EM modes tuned at the same frequency may enter the strong-coupling regime, manifested by the emergence of two hybrid light–matter modes. Certainly, there are numerous material combinations that allow such resonances to coexist. However, the key to achieving especially high coupling strengths lies in the combination of optical transitions characterized by large dipole moments with photonic cavities that tightly confine EM fields.

Composite structures consisting of metallic and excitonic components can be easily designed to exhibit both requirements. On the one hand, open plasmonic cavities are known to accomplish subwavelength field confinement and significant field enhancement, owing to the excitation of SPPs and LSP modes [1, 2]. On the other hand, J–aggregates in organic molecules have been shown to sustain transitions of very large oscillator strengths, and consequently strong dipole moments at room temperature [3, 4]. At the same time, advances in the design of high-quality low-dimensional semiconductors, like quantum dots [5] and two-dimensional TMDCs [6], that support excitons of large collective moments and narrow transition lines, have opened new possibilities for what can play the role of the QE. Yet, the Ohmic losses in metals oftentimes hinder the coherent energy exchange with the exciton. Recently, dielectric nanostructures have emerged as alternative low-loss photonic cavities [7]. Indeed, silicon NPs have been shown, both theoretically [8, 9] and experimentally [10, 11], to yield an appreciable coupling strength.

Whether it concerns plasmonic or Mie-resonant cavities, it has been envisaged that the mode hybridization achieved by these structures when combined with organic

P. E. Stamatopoulou, *Strong Light-Matter Interactions in Extreme Plasmonic and Mie-Resonant Systems*, Springer Theses, https://doi.org/10.1007/978-3-031-67907-0_3

molecules can be beneficial for chirality sensing [12]. Even in the weak coupling regime, metallic and dielectric NPs placed in close proximity to chiral molecules have been shown to drastically modify the chiroptical response of the system [13]. The next natural step is to explore what other effects could arise through the strong coupling of chiral light and matter, thus igniting the quest for chiral polaritons [14].

In this chapter we explore systems consisting of spherical NPs and assemblies thereof, coupled to excitonic materials. We first generalize Mie theory for NPs surrounded by a concentric spherical shell (*core–shell* NPs) and, subsequently, study the optical response of spherical metallic and dielectric resonators encapsulated in an excitonic medium. In the case of plasmon–exciton coupling, we briefly explore the emerging polaritonic modes beyond the LRA, as predicted by the nonlocal models described in Sect. 2.3.5. Finally, once we have consolidated the theoretical basis of strong coupling with simple isotropic NPs, we introduce chiral cavities and investigate the properties of chiral polaritons, following the work presented in Publication B.

3.1 Mie Theory for Core–Shell Nanoparticles

One advantage of the core–shell architecture for the study of strong light–matter interactions is that, for the theoretical description, we can employ Mie theory, which can be easily generalized for NPs consisting of an arbitrary number of shells concentric to the core [15]. In the spirit of standard Mie theory, the EM problem illustrated in Fig. 3.1 can be solved by expressing the fields in each layer of the system, including the host medium, as a linear combination of standing and outgoing spherical

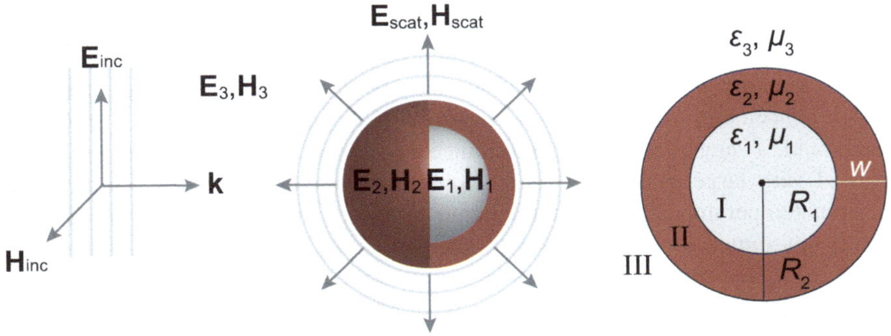

Fig. 3.1 Mie theory for core–shell NPs. A spherical NP of radius R_2, consisting of a core (region I) of radius R_1 and optical parameters (ε_1, μ_1) encapsulated in a shell (region II) of width w and parameters (ε_2, μ_2), is suspended in a host medium (region III) with (ε_3, μ_3), and illuminated by a plane EM wave $(\mathbf{E}_{\text{inc}}, \mathbf{H}_{\text{inc}})$. The incident wave is scattered on the surface of the NP in the form of outgoing spherical waves $(\mathbf{E}_{\text{scat}}, \mathbf{H}_{\text{scat}})$. The total field in region I $(\mathbf{E}_1, \mathbf{H}_1)$, takes the form of standing waves, whereas in regions II $(\mathbf{E}_2, \mathbf{H}_2)$ and III $(\mathbf{E}_3, \mathbf{H}_3)$ it consists of both standing and outward propagating spherical waves

waves of TE and TM type (for a detailed derivation see Appendix D.2). All unknown coefficients can then be found via the boundary conditions; here, the tangential components of the fields must be continuous at two interfaces, i.e. for $r = R_1$ (inner radius) and $r = R_2$ (outer shell radius). The resulting sets of linear equations give the Mie coefficients connecting the scattered fields to the incident, as

Box 3.1 | Mie coefficients for the scattered fields in core–shell NPs

$$T_{\mathrm{E}\ell} = \frac{\left[\varepsilon_2 h_\ell^+(x_2)\Psi_\ell'(x_3) - \varepsilon_3 \xi_\ell'(x_2) j_\ell(x_3)\right] T_{\mathrm{E}\ell}^{22} + \varepsilon_2 j_\ell(x_2)\Psi_\ell'(x_3) - \varepsilon_3 \Psi_\ell'(x_2) j_\ell(x_3)}{\left[\varepsilon_3 h_\ell^+(x_3)\xi_\ell'(x_2) - \varepsilon_2 \xi_\ell'(x_3) h_\ell^+(x_2)\right] T_{\mathrm{E}\ell}^{22} + \varepsilon_3 h_\ell^+(x_3)\Psi_\ell'(x_2) - \varepsilon_2 \xi_\ell'(x_3) j_\ell(x_2)}, \quad (3.1a)$$

$$T_{\mathrm{M}\ell} = \frac{\left[\mu_2 h_\ell^+(x_2)\Psi_\ell'(x_3) - \mu_3 \xi_\ell'(x_2) j_\ell(x_3)\right] T_{\mathrm{M}\ell}^{22} + \mu_2 j_\ell(x_2)\Psi_\ell'(x_3) - \mu_3 \Psi_\ell'(x_2) j_\ell(x_3)}{\left[\mu_3 h_\ell^+(x_3)\xi_\ell'(x_2) - \mu_2 \xi_\ell'(x_3) h_\ell^+(x_2)\right] T_{\mathrm{M}\ell}^{22} + \mu_3 h_\ell^+(x_3)\Psi_\ell'(x_2) - \mu_2 \xi_\ell'(x_3) j_\ell(x_2)}, \quad (3.1b)$$

where $x_i = k_i R_2$, for $i = 2, 3$ and $T_{\mathrm{E/M}\ell}^{22}$ refer to the Mie coefficients of the uncoated NP of Eqs. (2.77), evaluated at the boundary between regions I and II, namely at $r = R_1$. Similar exact expressions can be derived to connect the incident fields with the ones induced at any other layer (see Appendix D.2). Eventually, the optical cross sections can be calculated by Eqs. (2.78), as usual.

3.2 Strong Coupling of Plasmons with Excitons

Let us begin the discussion by considering the combination that is most commonly encountered in strong-coupling studies in the field of nanophotonics, the plasmon–exciton coupling. The quasiparticle arising from the coupling of these two excitations with light is a plasmon–exciton polariton of hybrid light–matter nature, also known as *plexciton* [17]. Here, we examine the plexcitonic resonances emerging in core–shell NPs, meaning that it is LSP modes that couple to the excitonic transition.

Consider a silver (Ag) NP of radius $R_1 = 5$ nm, and permittivity given by the Drude model of Eq. (2.56) with $\varepsilon_\infty = 3.3$, $\hbar\omega_\mathrm{p} = 9.17$ eV, $\hbar\gamma = 21$ meV, as seen in Fig. 3.2a. Due to its size, the NP almost exclusively absorbs the incident illumination, while scattering is negligible, featuring a sharp ED mode at $\hbar\omega_\mathrm{ED} = 3.98$ eV in Fig. 3.2d, in accord with the prediction of the quasistatic approximation of Eq. (2.81). Next, we introduce an excitonic shell of width $w = 2$ nm described by the dielectric function depicted in Fig. 3.2b, which follows the Lorentz model of Eq. (2.61), for parameter values $\varepsilon_\infty = 1$, $f = 0.03$, $\hbar\omega_\mathrm{exc} = 3.98$ eV, $\hbar\gamma_\mathrm{exc} = 21$ meV. The extinction cross section of the excitonic shell presented in Fig. 3.2e reveals a sharp peak at $\hbar\omega_\mathrm{exc}$ due to the exciton transition, as well as an additional (void) resonance, slightly blueshifted with respect to the former, which is related to the geometrical characteristics of the shell [18].

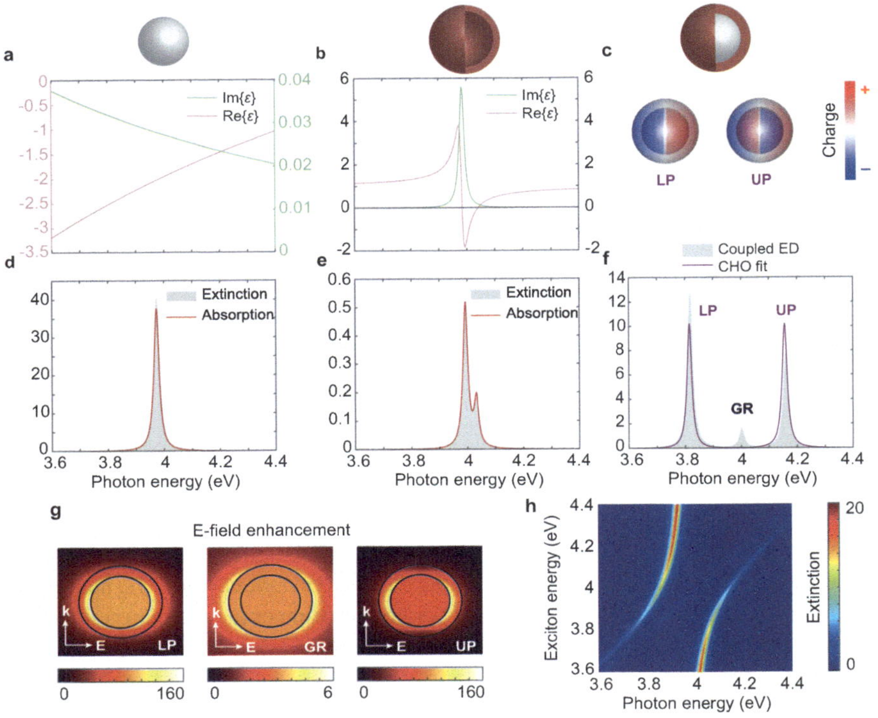

Fig. 3.2 Strong coupling in Ag NPs. a, b Real (pink curve) and imaginary part (green curve) of the dielectric function of **a** the Ag core, described by the Drude model of Eq. (2.56) with $\varepsilon_\infty = 3.3$, $\hbar\omega_p = 9.17$ eV, $\hbar\gamma = 21$ meV, and **b** the excitonic shell, described by the Lorentz model of Eq. (2.61) with $\varepsilon_\infty = 1$, $f = 0.03$, $\hbar\omega_{exc} = 3.98$ eV, $\hbar\gamma_{exc} = 21$ meV. **c** Charge distribution on the inner and outer surface of a core–shell NP consisting of an Ag core of radius $R_1 = 5$ nm encapsulated in a concentric shell of the excitonic medium of width $w = 2$ nm, calculated with the MNPBEM package at the energy of the LP and the UP [16]. **d, e, f** Normalized extinction (gray shaded area) and absorption (dark red solid curve) cross sections of **d** the Ag NP of radius $R_1 = 5$ nm, **e** the excitonic shell of inner radius $R_1 = 5$ nm and outer $R_2 = 7$ nm, and **f** the core–shell NP of panel **c**. The two polaritonic modes are fitted using the CHO model of Eq. (2.100) (purple line), and yield coupling strength $\hbar g = 340$ meV. **g** Enhancement of the total electric field evaluated at the energies of the LP, GR, UP modes. **h** Normalized extinction cross section as a function of the photon and the exciton energy of the coupled core–shell NP

Even though the extinction cross section of the excitonic shell is two orders of magnitude lower than that of the Ag core, once the two are combined in Fig. 3.2f we observe the emergence of two sharp peaks of comparable height (labeled LP and UP in panel f) at energies 3.82 and 4.16 eV, separated by energy difference $\hbar g = 340$ meV. The avoided crossing of the extinction peaks, formed when the plasmon and exciton resonances are detuned in Fig. 3.2h, is a first indication of their polaritonic nature. Strong coupling can be safely claimed considering the value of the energy separation compared to the damping rates of the original modes ($\hbar\gamma_{\mathrm{ED}} = \hbar\gamma_{\mathrm{exc}} = 21$ meV), which clearly satisfies the strong-coupling criterion presented in Box 2.15.

In Fig. 3.2f and h, we observe the emergence of a third peak in between the LP and UP modes, at 4 eV. Reference [18] provides a thorough investigation of the origin and behavior of this peak, and shows that it corresponds to a geometric resonance (GR) of the system responding almost as a uniform entity. This can be corroborated in Fig. 3.2g, where we calculate the electric field enhancement, and find a homogeneous profile at this energy. On the other hand, at the energies of the LP and UP modes we observe strong enhancement of the field at the inner surface, where the coupling takes place. Interestingly, the distribution of charges at the interfaces in Fig. 3.2c reveals that the LP mode corresponds to in-phase oscillating charges on the inner and outer surface of the NP, whereas at the UP mode they oscillate out of phase.

Naturally, metallic systems of characteristic sizes of a few nanometers, as the ones studied in this section, cannot be accurately described by LRA. Nevertheless, this simplistic approximation can capture qualitatively the most important aspects of their optical response, while also giving a rough estimate of the coupling strength which, as we see in the next section, is not significantly affected by nonlocal and finite-size effects.

3.2.1 Nonlocal Effects in Plexcitonic Systems

Let us now examine the influence of nonlocal corrections on strong coupling. As discussed in Sect. 2.3.5, nonlocality causes energy shifts and increased damping of the plasmon resonances, which manifest as spectral shifts and linewidth broadening of the corresponding extinction peaks. The key question to answer in the context of strong coupling is whether these changes will affect the energy split, and as a consequence the plexciton coupling strength.

Sticking to the same core–shell NP studied in the previous section, in Fig. 3.3a we plot the extinction of the Ag core of radius $R_1 = 5$ nm as reproduced by the nonlocal models described in Sect. 2.3.5, compared to the LRA prediction (gray shaded area), assuming a Fermi velocity equal to $v_{\mathrm{F}} = 1.39 \cdot 10^6$ m/s. The extinction cross section within the Kreibig SDB model [Eq. (2.118) with $A = 1$], that only accounts for a size-dependent damping, shows a linewidth broadening of the ED mode equal to $\hbar v_{\mathrm{F}}/R_1 \approx 183$ meV in comparison to LRA. This is, of course, not surprising, since it is exactly the correction to the Drude damping that we introduce by hand [see Eq. (2.118)]. All other models reproduce the spectral shift to the blue, observed in noble metals due to screening from d-band electrons. Besides, the increased screening is reflected in the relatively high value for $\varepsilon_\infty = 3.3$ we consider for Ag. In particular, considering parameters $\mathscr{b} = \sqrt{3/5}v_{\mathrm{F}}$ and $\mathscr{D} = 9.62 \cdot 10^{-4}$ m^2/s for silver [19], the HDM and GNOR spectra are blueshifted compared to the LRA prediction by 59 and 96 meV, respectively. In addition to the blueshift, GNOR also accounts for surface-enhanced Landau damping that manifests as resonance broadening, increased by 200 meV.

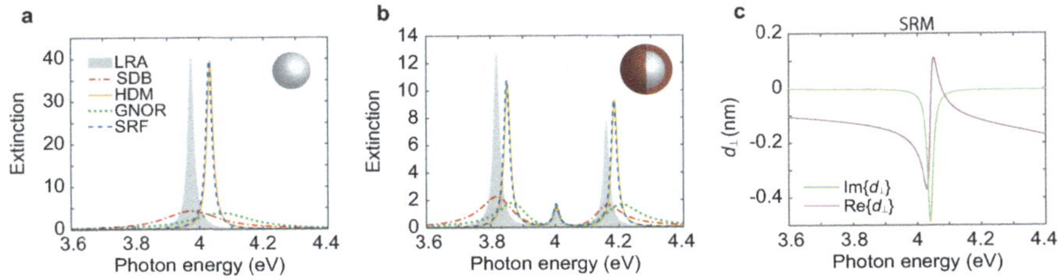

Fig. 3.3 Strong coupling in Ag NPs beyond LRA. **a** Normalized extinction cross section of an Ag NP of radius $R_1 = 5$ nm within the LRA (gray shaded area), Kreibig SDB model (red dashed-dotted line), HDM (solid yellow line), GNOR (green dotted line), and Feibelman SRF (blue dashed line) with $d_\parallel = 0$, and $d_\perp$ given by the SRM model, as plotted in panel **c**. **b** Same as in **a**, for an Ag NP covered by an excitonic shell of width $w = 2$ nm, and parameters $\varepsilon_\infty = 1$, $f = 0.03$, $\hbar\omega_{\text{exc}} = 3.98$ eV and $\hbar\gamma_{\text{exc}} = 21$ meV in Eq. (2.61). **c** The real and imaginary part of $d_\perp$ obtained for Ag with Eq. (3.3), using the specular-reflection model (SRM)

For the calculation of extinction within SRF, we use the d-parameter values obtained for Ag in the specular-reflection model (SRM) [20]. In this model, the medium is treated as an electron gas in which the electrons are considered to be specularly reflected at the surface of the metal. Thereby the electron density is eliminated beyond the assumed surface. The nonlocal response of the metal is encoded in its k-dependent bulk permittivity, and can be incorporated in the d-parameters via the expression [21]

$$d_\perp^{\text{SRM}} = -\frac{2}{\pi}\frac{\varepsilon_1\varepsilon_2}{\varepsilon_1 - \varepsilon_2}\int_0^\infty \frac{dk}{k^2}\left(\frac{1}{\varepsilon_\parallel(k,\omega)} - \frac{1}{\varepsilon_1}\right), \tag{3.2}$$

where $\varepsilon_\parallel$ is the longitudinal component of the permittivity. For a dielectric function given by the HDM model of Eq. (2.68b), the integral yields

$$d_\perp^{\text{SRM}}(\omega) = \frac{i\varepsilon_2\varepsilon_1}{\varepsilon_1 - \varepsilon_2}\frac{\beta}{\omega_\text{p}\sqrt{\varepsilon_\infty}}\left[\frac{\varepsilon_\infty}{\varepsilon_1} - 1\right]^{3/2}. \tag{3.3}$$

Figure 3.3c shows the values of $d_\perp^{\text{SRM}}$ obtained for Ag using Eq. (3.3). Using these values, and for $d_\parallel = 0$ due to charge neutrality of the surface, we find in Fig. 3.3a an extinction cross section similar to the one reproduced by HDM, namely a blueshift of 56 meV and negligible broadening ($\sim 10^{-4}$).

The Ag core–excitonic shell coupled system presented in Fig. 3.3b exhibits the UP and LP branches. In all models the spectral shift and broadening of both modes follows the same trend as in the uncoupled NP, while the geometric resonance remains unaffected. This could be anticipated, as the latter feature originates from the excitonic material exclusively and does not participate in the coupling. Since both plexcitonic modes blueshift by approximately the same energy, overall the same energy split is retrieved by all models presented in Fig. 3.3b, and thus the coupling remains unaffected by the incorporation of nonlocal and finite-size effects. When compared

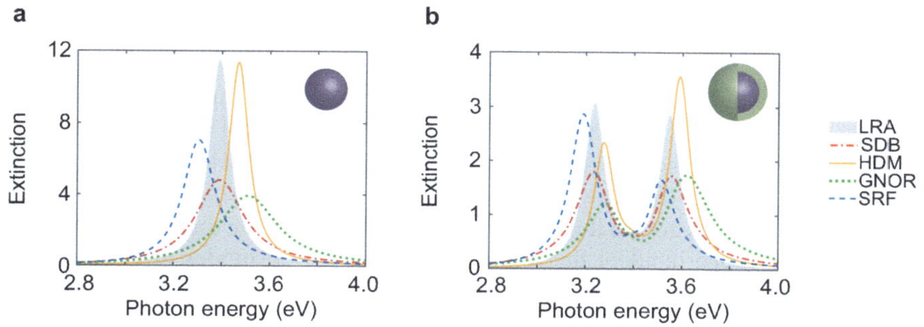

Fig. 3.4 Strong coupling in Na NPs beyond LRA. **a** Normalized extinction cross section of a Na NP of radius $R_1 = 5$ nm within the LRA (gray shaded area), Kreibig SDB model (red dashed-dotted line), HDM (solid yellow line), GNOR (green dotted line), and Feibelman SRF (blue dashed curve) with $d_\perp$ given by the values calculated in Ref. [23], and for $d_\parallel = 0$. **b** Same as in **a**, for a Na NP covered by an excitonic shell of width $w = 2$ nm, and parameters $\varepsilon_\infty = 1$, $f = 0.02$, $\hbar\omega_{\rm exc} = 3.38$ eV and $\hbar\gamma_{\rm exc} = 100$ meV in Eq. (2.61)

to the SDB and GNOR models that predict a significant increase in damping, exhibiting linewidths $\hbar\gamma_{\rm SDB} = 205$ meV and $\hbar\gamma_{\rm GNOR} = 223$ meV, respectively, the energy split (340 meV) still satisfies the strong coupling criterion. The robustness of strong coupling as predicted by GNOR has been reported before for silver NPs [22].

Figure 3.4 shows the extinction cross section of a Na NP of radius $R = 5$ nm as predicted by the various nonlocal models compared to LRA. Due to its low work function, Na exhibits spill-out, which is expected to produce a redshift of the plasmonic modes sustained in the metal. However, as suggested by Table 2.1, the trend of the spectral shift can only be captured by the Feibelman framework, since all other models are not designed to reproduce spill-out effects. Indeed Fig. 3.4a shows that the SRF model, employing the d-parameters calculated in Ref. [23] and plotted in Fig. 2.10c, results in a redshift of the plasmonic mode from 3.39 to 3.31 eV, whereas SDB, HDM and GNOR wrongly produce a blueshift. Once the Na NP is coupled to an excitonic shell, LRA predicts an energy splitting of 312 meV, as illustrated in Fig. 3.4b. In the same panel, we observe that in the SRF model the UP and LP branches are redshifted equally, leading to negligible changes to the coupling strength, of the order of 10 meV ($\hbar g = 322$ meV). Publication G offers a rigorous study of the coupling strength obtained in sodium–excitonic core–shell NPs, and compares the results acquired in LRA and the Feibelman approach, showcasing negligible disagreement. We can therefore conclude that the LRA model yields the correct value of the coupling strength even for dimensions below 10 nm, as long as Landau damping is not large enough to prevent the energy exchange between the coupled components.

3.3 Strong Coupling of Dielectric Mie Modes with Excitons

Admittedly, plasmons offer the best field confinement, a key ingredient to achieve high coupling strengths. Nevertheless, they come with the cost of high Ohmic losses. In order to deal with the inevitable loss in plasmonics, dielectric resonators, most often made of silicon (Si), are explored as an alternative. Even though dielectric cavities do not offer as good a field confinement as plasmons, the realization that they significantly enhance incident fields and support long-lived modes of both electric and magnetic multipolar character, has opened new opportunities in the emerging field of all-dielectric photonics. In recent years, the possibility of achieving strong coupling between excitons and Mie resonances has been confirmed both theoretically and experimentally [8–10, 24]. In the spirit of these works, we study here the coupling between the Mie resonances excited in a Si NP and the excitonic transitions supported by the material in which the Si sphere is encapsulated.

In Fig. 3.5d and g we show the optical cross sections and the contributions of the first few multipoles to the extinction for a Si NP of radius $R = 85$ nm, in an energy window in which Si exhibits high positive values of $\mathrm{Re}\{\varepsilon\}$ and very low losses (see Fig. 3.5a). In order to couple the NP to an excitonic resonance, we design a shell that supports an optical transition of energy similar to that of the MD Mie mode, lying at 1.83 eV. This requirement is fulfilled by the Lorentz dielectric function of Eq. (2.61), plotted in Fig. 3.5b for parameter values $\varepsilon_\infty = 3$, $f = 0.2$, $\hbar\omega_{\mathrm{exc}} = 1.77$ eV, $\hbar\gamma_{\mathrm{exc}} = 0.05$ eV. Figure 3.5e and h present the extinction cross section of the shell, together with its decomposition into absorption and scattering (panel e), and into electric and magnetic multipoles (panel h). A noteworthy comment here is that, when choosing the value for the transition energy, one has to take into account the background permittivity of the Lorentz function. In our example, in the absence of the excitonic resonance the coated NP is embedded in a host medium of effective permittivity $3\varepsilon_0$, resulting in a smaller contrast between the permittivities of the two media, and as a consequence an overall redshift of the Si spectrum.

Figure 3.5c, f and i show the splitting of the MD mode, that can be observed in both scattering and absorption spectra. It is, however, barely discernible in the latter, because of the overall very low and broad absorption peak of the Si NP at this energy. Already the splitting of the MD mode, and its anticrossing trend as we detune the photon and the exciton energies in Fig. 3.5, are good indications that the system operates in the strong coupling regime. As a final step, one can resort to the strong coupling criterion of the CHO model given in Box 2.15.

For dielectric NPs like Si, that feature numerous overlapping asymmetric modes, a rigorous fitting would require a model of multiple harmonic oscillators. To avoid suffering a very cumbersome calculation for the purpose of applying a rather empirical criterion, in the CHO model of Fig. 3.6 we consider the coupling of the excitonic transition with the MD mode alone as emerging in a NP coated by a shell with $\varepsilon_2 = 3$, and disregard the contribution of the ED tail. As a first step, we fit the MD mode of the coated uncoupled Si NP (i.e. in the absence of excitonic transitions in the shell) with two Lorentzian curves to account for the asymmetry of the mode; one captures

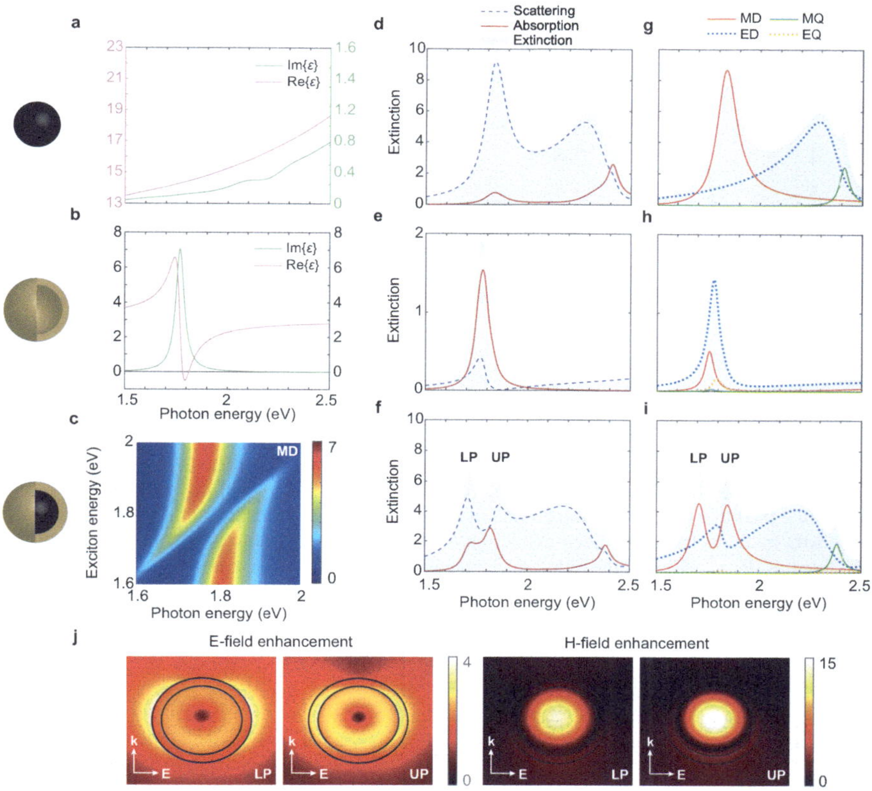

Fig. 3.5 **Strong coupling in Si NPs**. **a**, **b** Real and imaginary part of the dielectric function of **a** Si according to Ref. [25], and **b** the excitonic material described by the Lorentz dielectric function of Eq. (2.61) with parameters $\varepsilon_\infty = 3$, $f = 0.2$, $\hbar\omega_{\text{exc}} = 1.77$ eV, $\hbar\gamma_{\text{exc}} = 0.05$ eV. **d**, **e**, **f** Normalized extinction (gray shaded area), absorption (dark red solid curve) and scattering (dark blue dashed curve) cross sections of **d** a Si NP of radius $R_1 = 85$ nm, **e** a shell of the excitonic material of **b** of inner radius $R_1 = 85$ nm and outer $R_2 = 105$ nm, and **f** the Si NP coated with the excitonic shell. **g**, **h**, **i** Normalized extinction cross section and modal decomposition into the MD (red solid line), MQ (green solid line), ED (blue dotted line) and EQ mode (yellow dotted line) of **g** the Si NP, **h** the excitonic shell, and **i** the Si NP encapsulated in the exciton shell. The coupled system exhibits a splitting of the MD mode that follows an anticrossing trend plotted in panel **c** for varying photon and exciton energy. **j** Enhancement of the total electric (left) and magnetic field (right) evaluated at the energies of the LP and UP modes

the peak, while the other the shoulder of the asymmetric curve. From the fitting we extract the energy and damping rate of the mode, and obtain $\hbar_{\text{MD}} = 1.78$ eV and $\hbar\gamma_{\text{MD}} = 0.92$ eV, respectively. We then fit the MD contribution to the extinction of the coupled system (red solid curve in Fig. 3.5i) with the extinction cross section within the CHO model of Eq. (2.100) (black dotted curve) to obtain a rough estimate of the coupling strength, which yields $\hbar g = 137$ meV. This value satisfies the strong-coupling criterion

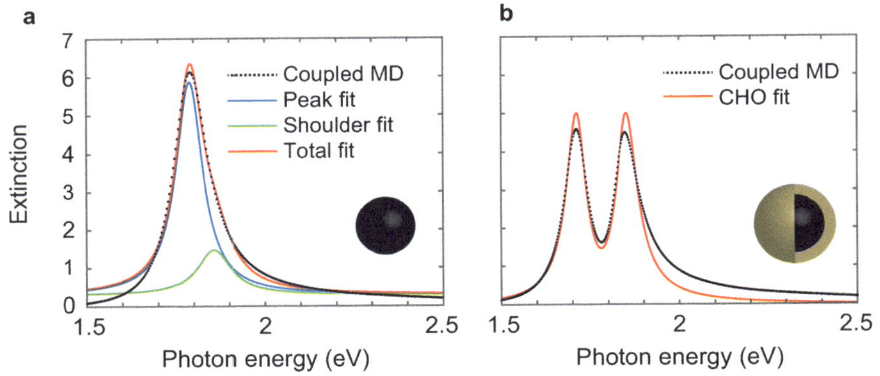

Fig. 3.6 CHO model for Si core—shell NPs. **a** Contribution of the MD mode to the normalized extinction cross section (black dotted line) of the uncoupled Si NP of radius $R_1 = 85$ nm, embedded in a concentric shell of dielectric constant $\varepsilon_2 = 3$ and width $w = 20$ nm. The asymmetric lineshape can be well fitted by the sum of two Lorentzian curves (red solid line), one fitting the peak (blue solid line), while the other reproduces the broad tail at higher energies (green solid line). **b** Contribution of the MD mode to the normalized extinction cross section (black dotted line) of the coupled Si NP of radius $R_1 = 85$ nm, embedded in a concentric shell of the excitonic material depicted in Fig. 3.5b, of width $w = 20$ nm, and fitted with the respective extinction within the CHO model (red solid line)

$$\hbar g = 137 \text{ meV} > \sqrt{\frac{\hbar\gamma_{\text{MD}}^2 + \hbar\gamma_{\text{exc}}^2}{2}} = \sqrt{\frac{92^2 + 50^2}{2}} = 74 \text{ meV} . \tag{3.4}$$

We have therefore shown that strong coupling is also possible between exciton and Mie resonances in dielectric cavities.

3.4 Mie Theory for Chiral Nanoparticles

Ever since the realization that most biological molecules are chiral, their interaction with circularly polarized light has been a subject of particular interest. Chiral molecules appear in two enantiomeric versions of degenerate transition frequencies. Due to the degeneracy, common spectroscopic techniques are not able to distinguish between the two configurations. In order to identify the enantiomers and quantify the chiral properties of a medium, CD spectroscopy is typically employed. In this technique the molecules are excited by L/RCP light, which breaks the degeneracy, and reveals the distinctive response of each enantiomeric configuration. However, conventional CD is restricted to a narrow range of applications, due to the general weakness of the corresponding signals in natural chiral materials [26].

Recently, several approaches for enhancing the sensitivity of the measurements have been proposed that take advantage of the strong near-field enhancement provided by resonant single NPs and NP arrays [27–29] or self-assemblies [30]. The direction of CD enhancement via coupling of the chiral medium to other resonant objects has led to the conceptualization of chiral polaritons. Here, we attempt to understand and explore the chiroptical properties of such exotic structures, making use of the concept of strong coupling of nanoresonators encapsulated in excitonic media, as described in the previous sections.

Mie theory can easily be generalized to encompass chiral media, described by the constitutive relations of Eqs. (2.41)

$$\mathbf{D} = \varepsilon\varepsilon_0 \mathbf{E}(\mathbf{r}, \omega) + \frac{i\kappa}{c}\mathbf{H}\,, \tag{3.5a}$$

$$\mathbf{B} = -\frac{i\kappa}{c}\mathbf{E} + \mu\mu_0 \mathbf{H}\,. \tag{3.5b}$$

Since only circularly polarized light can travel inside chiral media unaffected, it is convenient to express the fields inside the NP as a combination of LCP and RCP light waves characterized by wavenumbers $k_{\mathrm{L/R}} = k_1 \pm \kappa\omega/c$, yielding (see Appendix D.3)

$$\begin{aligned}\mathbf{E}_{\mathrm{in}}(\mathbf{r}) = \sum_{\ell m} \Bigg\{ & c^{\mathrm{L}}_{\ell m}\left[j_\ell(k_{\mathrm{L}} r)\mathbf{X}_{\ell m}(\hat{\mathbf{r}}) + \frac{1}{k_{\mathrm{L}}}\nabla \times j_\ell(k_{\mathrm{L}} r)\mathbf{X}_{\ell m}(\hat{\mathbf{r}})\right] \\ & + c^{\mathrm{R}}_{\ell m}\left[j_\ell(k_{\mathrm{R}} r)\mathbf{X}_{\ell m}(\hat{\mathbf{r}}) - \frac{1}{k_{\mathrm{R}}}\nabla \times j_\ell(k_{\mathrm{R}} r)\mathbf{X}_{\ell m}(\hat{\mathbf{r}})\right]\Bigg\}\,, \end{aligned} \tag{3.6a}$$

$$\begin{aligned}\mathbf{H}_{\mathrm{in}}(\mathbf{r}) = \frac{i}{Z_1}\sum_{\ell m} \Bigg\{ & - c^{\mathrm{L}}_{\ell m}\left[j_\ell(k_{\mathrm{L}} r)\mathbf{X}_{\ell m}(\hat{\mathbf{r}}) + \frac{1}{k_{\mathrm{L}}}\nabla \times j_\ell(k_{\mathrm{L}} r)\mathbf{X}_{\ell m}(\hat{\mathbf{r}})\right] \\ & + c^{\mathrm{R}}_{\ell m}\left[j_\ell(k_{\mathrm{R}} r)\mathbf{X}_{\ell m}(\hat{\mathbf{r}}) - \frac{i}{k_{\mathrm{R}}}\nabla \times j_\ell(k_{\mathrm{R}} r)\mathbf{X}_{\ell m}(\hat{\mathbf{r}})\right]\Bigg\}\,, \end{aligned} \tag{3.6b}$$

whereas the fields in the surrounding medium are the same as in the isotropic case. Following, once more, the steps of Mie theory, we derive the formulas for the optical cross sections, as

Box 3.2 | Optical cross sections in chiral media

$$\begin{aligned}\sigma^{\mathrm{L/R}}_{\mathrm{scat}} = \frac{2}{(k_2 R)^2}\sum_\ell (2\ell+1)\Big[& |T_{\mathrm{E}\ell}|^2 + |T_{\mathrm{M}\ell}|^2 + 2|T_{\mathrm{C}\ell}|^2 \\ & \mp 2\mathrm{Im}\Big\{T^*_{\mathrm{C}\ell}\,(T_{\mathrm{E}\ell} + T_{\mathrm{M}\ell})\Big\}\Big]\,, \end{aligned} \tag{3.7a}$$

$$\sigma_{\text{ext}}^{\text{L/R}} = -\frac{2}{(k_2R)^2}\sum_{\ell}(2\ell+1)\text{Re}\Big\{T_{\text{E}\ell}+T_{\text{M}\ell}\mp 2iT_{\text{C}\ell}\Big\}\,, \tag{3.7b}$$

$$\sigma_{\text{abs}}^{\text{L/R}} = \sigma_{\text{ext}}^{\text{L/R}} - \sigma_{\text{scat}}^{\text{L/R}}\,. \tag{3.7c}$$

In Eq. (3.7a) the Mie coefficients $T_{\text{E}\ell}$, $T_{\text{M}\ell}$ and $T_{\text{C}\ell}$ now read

$$T_{\text{E}\ell} = \frac{A(\text{L})V(\text{R})+A(\text{R})V(\text{L})}{W(\text{L})V(\text{R})+W(\text{R})V(\text{L})}\,, \tag{3.8a}$$

$$T_{\text{M}\ell} = \frac{B(\text{L})W(\text{R})+B(\text{R})W(\text{L})}{W(\text{L})V(\text{R})+W(\text{R})V(\text{L})}\,, \tag{3.8b}$$

$$T_{\text{C}\ell} = i\,\frac{A(\text{L})W(\text{R})-A(\text{R})W(\text{L})}{W(\text{L})V(\text{R})+W(\text{R})V(\text{L})}\,, \tag{3.8c}$$

where for $\text{P} = \text{L}, \text{R}$ we have set

$$A(\text{P}) = \frac{1}{Z_1k_2}j_\ell(k_\text{P}R)\Psi_\ell'(k_2R) - \frac{1}{Z_2k_\text{P}}\Psi_\ell'(k_\text{P}R)j_\ell(k_2R)\,, \tag{3.9a}$$

$$B(\text{P}) = \frac{1}{Z_2k_2}j_\ell(k_\text{P}R)\Psi_\ell'(k_2R) - \frac{1}{Z_1k_\text{P}}\Psi_\ell'(k_\text{P}R)j_\ell(k_2R)\,, \tag{3.9b}$$

$$V(\text{P}) = \frac{1}{Z_1k_\text{P}}h_\ell^+(k_2R)\Psi_\ell'(k_\text{P}R) - \frac{1}{Z_2k_2}\xi_\ell'(k_2R)j_\ell(k_\text{P}R)\,, \tag{3.9c}$$

$$W(\text{P}) = \frac{1}{Z_2k_\text{P}}h_\ell^+(k_2R)\Psi_\ell'(k_\text{P}R) - \frac{1}{Z_1k_2}\xi_\ell'(k_2R)j_\ell(k_\text{P}R)\,. \tag{3.9d}$$

In Eqs. (3.8), $T_{\text{C}\ell}$ is a cross-term element, linking the amplitude of the incident wave of TE/TM polarization with the amplitude of the scattered wave of opposite polarization (TM/TE).

If we define CD as the difference between the absorption cross sections for LCP and RCP light excitation, utilizing the expressions in Box 3.2 we obtain a closed-form expression

$$\text{CD} = \sigma_{\text{abs}}^{\text{L}} - \sigma_{\text{abs}}^{\text{R}} = \frac{8}{(k_2R)^2}\sum_{\ell}(2\ell+1)\text{Im}\Big\{T_{\text{C}\ell}^*\left(T_{\text{E}\ell}+T_{\text{M}\ell}+1\right)\Big\}\,. \tag{3.10}$$

One may also define the differential scattering (DS) as the difference in scattering cross section of LCP and RCP waves

$$\text{DS} = \sigma_{\text{scat}}^{\text{L}} - \sigma_{\text{scat}}^{\text{R}} = -\frac{8}{(k_2R)^2}\sum_{\ell}(2\ell+1)\text{Im}\Big\{T_{\text{C}\ell}^*\left(T_{\text{E}\ell}+T_{\text{M}\ell}\right)\Big\}\,. \tag{3.11}$$

We note that one can introduce additional concentric layers on top of the chiral NP and derive the respective optical cross sections by following a methodology similar to the one presented in Sect. 3.1. In this manner we calculate the chiroptical response of the chiral core–shell structures presented in the following section.

3.4.1 Strong Coupling of Chiral Nanoparticles with Excitons

Let us start by revisiting the local chirality density of Eq. (2.45) in a lossless and non-disperive medium, given by

$$C(\mathbf{r}, \omega) = \frac{\varepsilon\omega}{2c^2}\mathrm{Im}\Big\{\mathbf{E}(\mathbf{r}, \omega) \cdot \mathbf{H}^*(\mathbf{r}, \omega)\Big\}. \tag{3.12}$$

Essentially, Eq. (3.12) associates the chirality density with the existence of parallel electric and magnetic field components at a point $\mathbf{r}$ in space that are properly phase-shifted, so that their inner product remains non-zero. To achieve large values of chirality density, we require high near-field intensities in confined volumes. Even though enhanced sensing of chirality is not in the scope of this thesis, it is important to note that, for efficient CD enhancement, optical cavities of pure handedness are highly desirable [29, 31]. It has been shown that this condition is met in metal or dielectric nanostructures which sustain overlapping modes of electric and magnetic type [32].

Taking the above discussion into account, we consider a NP reminiscent of the one studied in Sect. 3.3, which exhibits a considerable overlap between its ED and MD modes. Assume a sphere of radius $R = 85$ nm, characterized by a non-dispersive dielectric function, $\varepsilon_1 = 12$, permeability $\mu = 1$, and negligible loss, in order to keep complexity at a minimum. Figure 3.7a shows that, as expected by the real ε, the incident light only dissipates through the process of scattering. Indeed, Fig. 3.7d reveals a significant overlap between the ED and the MD mode, at around 2 eV, as well as with the MQ above 2.8 eV. As a result, we expect hotspots of high chirality density around and inside the NP[1] at these energies. To examine our claim, in Fig. 3.7g we plot the average chirality density enhancement $\langle C\rangle$, defined as the local chirality density integrated across the volume of the NP

$$\langle C(\omega)\rangle = \frac{3}{4\pi R^3}\int_0^R drr^2 \int d\Omega \, \frac{C(\mathbf{r}, \omega)}{\left|C^0_{\mathrm{L/RCP}}(\omega)\right|}. \tag{3.13}$$

Here $C^0_{\mathrm{L/RCP}} = \pm\omega\varepsilon_0/(2c)\,|E_0|^2$ is the chirality density of incident L/RCP light of amplitude E_0 propagating in vacuum. Indeed, close to the energies of significant mode overlap we observe sharp peaks, which are an indication of hotspots in the local chirality density in the volume of the structure. However, since C takes both negative and positive values, regions of high chirality density of opposite sign may

[1] The extinction cross section of the modes shown in Fig. 3.7d relates the scattered field to the incident field, meaning that it contains information only about the fields surrounding the sphere. However, the Lorentzian curves emerge from the poles of the Mie coefficients, that lie at the same energies as the ones connecting the internal field of the NP with the incident [see Eqs. (D.18)].

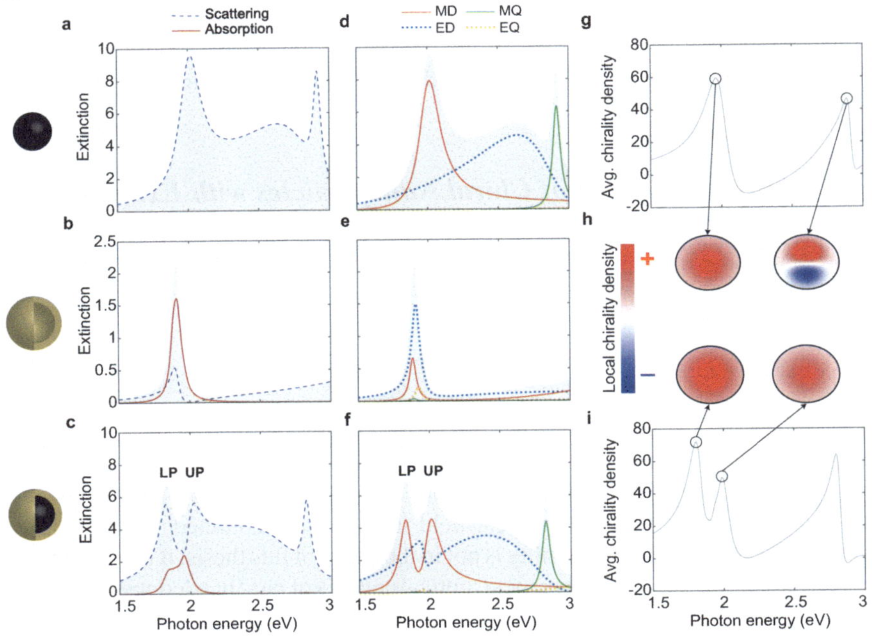

Fig. 3.7 Strong coupling of an achiral NP and an excitonic shell. a, b, c Normalized extinction (gray shaded area), absorption (dark red solid curve) and scattering (dark blue dashed curve) cross sections of **a** a NP of radius $R_1 = 85$ nm and $\varepsilon = 12$, **b** an excitonic shell of inner radius $R_1 = 85$ nm and outer $R_2 = 105$ nm, and permittivity given by Eq. (2.61) (with $\varepsilon_\infty = 3$, $f = 0.2$, $\hbar\omega_{\text{exc}} = 1.9$ eV, $\hbar\gamma_{\text{exc}} = 0.05$ eV), **c** the NP coated with the excitonic shell. **d, e, f** Normalized extinction cross section and modal decomposition into the MD (red solid line), MQ (green solid line), ED (blue dotted line) and EQ mode (yellow dotted line) of **d** the uncoupled NP, **e** the excitonic shell, and **f** the coupled system. **g** Average chirality density integrated over the volume of the dielectric NP surrounded by air. **h** Local chirality density in the volume of the dielectric NP evaluated at the energy of the first ($\hbar\omega = 1.96$ eV), and the second peak ($\hbar\omega = 2.89$ eV). **i** Average chirality density integrated over the volume of the dielectric NP, encapsulated in the excitonic shell and surrounded by air. In all panels we consider LCP illumination

arise at any energy. Nonetheless, the local chirality density evaluated at the energy of the first peak at ~1.96 eV is purely positive throughout the volume, as shown in Fig. 3.7h.

To explore the effect of an excitonic environment on the chiroptical response of optically active matter, it is important to first understand the impact of the excitonic coating on its resonant features, i.e. the overlapping electric and magnetic modes. We thus stick to the achiral NP, and introduce an exciton shell of width $w = 20$ nm, tuned at the energy of the MD mode of the dielectric NP, as presented in Fig. 3.7b and e [$\varepsilon_\infty = 3$, $f = 0.2$, $\hbar\omega_{\text{exc}} = 1.9$ eV, $\hbar\gamma_{\text{exc}} = 0.05$ eV in Eq. (2.61)]. The combined core–shell configuration is strongly coupled, exhibiting a mode splitting of the order

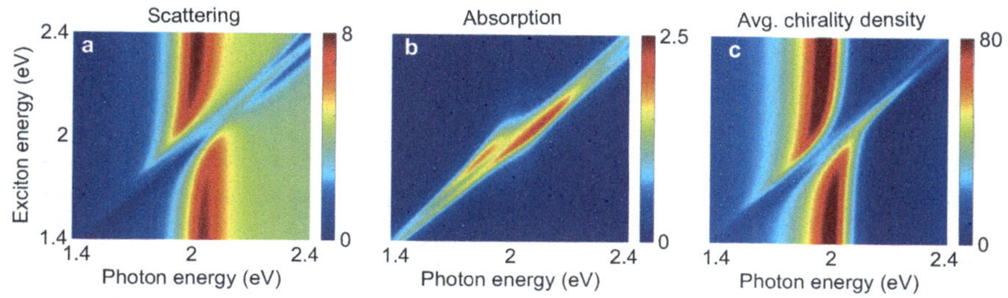

Fig. 3.8 Strong coupling of the chiral NP and the excitonic shell. a Normalized scattering cross section and, **b** normalized absorption cross section of the dielectric NP with chirality parameter $\kappa = 10^{-3}$, versus photon and excitonic transition energy. **c** Average chirality density integrated over the volume of the achiral dielectric NP (core), versus photon and excitonic transition energy

of 180 meV.[2] We observe that the average chirality density (integrated over the volume of the core) also exhibits a splitting of the resonant feature. Therefore, the hybridization of the MD mode due to the excitonic transition leads to the creation of two energy bands of high chirality density of the same sign.

We next explore the impact of the excitonic environment on an optically active configuration. Figure 3.8a and b show the scattering and absorption cross section of the same coated dielectric NP, now with a chiral core characterized by a Pasteur parameter $\kappa = 10^{-3}$ as we detune the system.[3] This particular value of the chirality parameter serves two purposes: on the one hand, it is small enough to ensure that the electric and magnetic modes will be coupled but will otherwise remain unaffected, and, on the other hand, it is real-valued, leading to a zero absorption of the core alone. Therefore the non-zero absorption seen in Fig. 3.8b stems from the absorptive shell. Even though the core is not absorbing, the coupled system exhibits a significant absorption component, stemming from the excitonic shell, and the corresponding peak in Fig. 3.8b, naturally, follows the exitonic transition energy.

In Fig. 3.9 we calculate the DS and CD spectra of the chiral NP coupled to the excitonic shell, for a real and an imaginary value of κ. In the DS spectra, in both cases, the peak lying at 1.9 eV in the spectrum of the uncoupled NP (orange curve) is split into two peaks once the excitonic shell is introduced (blue curve), as illustrated in Fig. 3.9a and b. As we detune the exciton transition energy in Fig. 3.9e and f, the DS spectra follow the anticrossing trend of the scattering spectrum of the uncoupled NP (Fig. 3.8a).

Assuming a real κ, the optically active core supports neither chiral nor achiral absorption, therefore leading to a zero CD signal in Fig. 3.9c (orange curve). One would expect a similar outcome once the core is encapsulated in the excitonic shell,

[2] We do not show here the CHO model, since the system is very similar to the one presented in Fig. 3.6, and we follow exactly the same prescription.

[3] We mention that the spectra presented in Fig. 3.8a and b look identical for an imaginary Pasteur parameter $\kappa = 10^{-3}i$.

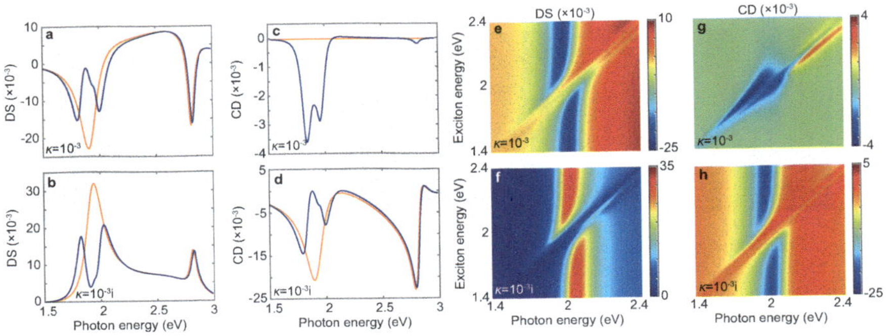

Fig. 3.9 DS and CD of the chiral NP coupled to the excitonic shell. a, b DS of the uncoupled (orange curve) and coupled (blue curve) NP. **c, d** CD of the uncoupled (orange curve) and coupled (blue curve) NP. **e, f** Photon versus exciton energy colour map of the DS corresponding to the coupled NP. **g, h** Photon versus exciton energy colour map of the CD corresponding to the coupled NP. In panels **a**, **c**, **e**, **g** we consider a real $\kappa = 10^{-3}$, whereas in **b**, **d**, **f**, **h** an imaginary $\kappa = 10^{-3}i$

since it does not possess chiral properties. In contrast, in Fig. 3.9c we find a significant CD component (blue curve), even comparable to the DS component. The reason for this is the fact that the chirality lifts the degeneracy of the Mie modes between excitation with LCP and RCP waves, as is already evident in the DS calculations. As a result, and despite the absence of absorbing chiral regions, asymmetric absorption takes place in the achiral regions of the combined shell–sphere system, mainly in the excitonic shell, thus leading to nonzero CD signal. Additionally, this is the reason why, eventually, the CD signal in the colormap of Fig. 3.9g inherits the behaviour of the absorption in the achiral NP, as presented in Fig. 3.8b. If instead we consider a core with imaginary $\kappa = 10^{-3}i$, in which case the uncoated NP already exhibits a significant CD signal (see orange curve in Fig. 3.9d), the CD spectrum attains a completely different behavior. What we observe in this case is the splitting of the spectral feature at 1.9 eV in panel d. Here, chiral absorption takes over; the contribution of the chiral material regions to asymmetric absorption dominates over that in the achiral regions, and CD locally behaves as predicted by Eq. (2.47), i.e. as $\text{CD} \propto \text{Im}(\kappa)C$, and therefore inherits the anticrossing trend of the chirality density C in Fig. 3.8c, as seen in Fig. 3.9h. To conclude, taking into account that realistic chiral media are characterized by complex Pasteur parameters, the characteristic footprint of strong coupling, i.e. the peak splitting and branch anticrossing, is always present in the relevant observables.

We have established so far that coupling with excitonic materials can reconfigure the chirality response of a system, even in the least sophisticated structures. We now move on to a structure that exhibits configurational rather than constitutional chirality. To this end, we consider a helix consisting of 7 Ag nanospheres (described by the experimental permittivity of Johnson and Christy [33]) of radius $R = 5$ nm, arranged

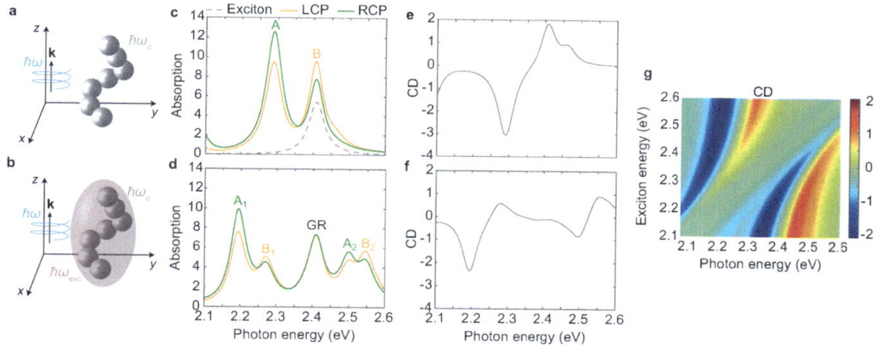

Fig. 3.10 **Chiral plasmon–exciton polaritons.** **a, b** Schematic illustration of an Ag-NP helix: 7 nanospheres, of radius $R = 5$ nm each, revolve around a hypothetical supporting pillar along the z axis, with steps of 30°, separated by a vertical distance $2R$ along z. **a** suspended in air, and **b** encapsulated within an excitonic matrix and surrounded by air. The helix is illuminated by circularly polarized light of frequency ω propagating along z, and excites a collective chain mode at energy $\hbar\omega_c$. **c** Normalized (to πR^2) absorption cross section of the uncoupled helix of **a**, for RCP (green curve) and LCP (yellow curve) incident light propagating along the helix axis, in the absence of an excitonic shell. The pink dashed curve corresponds to the absorption spectrum of the excitonic shell alone, suspended in air. The dielectric function of the excitonic material is given by Eq. (2.61) with $\varepsilon_\infty = 1$, $f = 0.02$, $\hbar\omega_{\text{exc}} = 2.4$ eV, $\hbar\gamma_{\text{exc}} = 0.052$ eV. **d** Normalized (to πR^2) absorption cross section of the helix embedded in the excitonic matrix as illustrated in panel **b**. **e** CD spectrum of the uncoupled helix of panel **a**. **f** CD spectrum of the coupled system of panel **b**. **g** Photon versus exciton energy color map of the CD corresponding to the coupled system

in a helical geometry, as shown in the schematics of Fig. 3.10a.[4] Such architectures are commonly fabricated with the DNA origami technique [34]. The NPs revolve around the z axis by steps of 30°, so that they constitute a full rotation, with a center-to-center distance of the spheres along z equal to $2R$. In this arrangement, neighboring NPs are nearly touching, thus ensuring intense plasmonic interactions, which are expected to lead to collective plasmonic chain modes [35]. Under L/RCP illumination, the absorption spectrum of Fig. 3.10c shows the emergence of two chain modes (labeled as A and B) at energies $\hbar\omega_c = 2.3$ and 2.4 eV, due to sub-chains excited along the two perpendicular polarization directions of the incident wave (here along the x and y axis), with slightly smaller energies because of the different effective chain length in each direction [35]. Interestingly, the two chain modes exhibit a different preference in absorption at the energies of the two resonances; as indicated by the color of the labels, resonance A absorbs RCP light more efficiently than LCP, and the opposite occurs for resonance B. The resulting CD spectrum in Fig. 3.10e features two peaks of opposite sign at the corresponding energies with an inflection point between them.

[4] The results presented in Fig. 3.10 were produced by Christos Tserkezis in COMSOL and they are presented in this thesis with his permission.

We introduce an excitonic environment around the helix, as illustrated in Fig. 3.10b, in the form of an ellipsoid just enclosing the entire helix (long axis 30 nm, short axis 17.5 nm). The matrix is described by the Lorentz permittivity according to Eq. (2.61), with $\varepsilon_b = 1$, $f = 0.02$, $\hbar\gamma_{\rm exc} = 0.052$ eV, and $\hbar\omega_{\rm exc} = 2.4$ eV, designed so that it features a resonance in its absorption at the same energy as chain mode B (see the absorption cross section of the self-standing excitonic matrix in Fig. 3.10c plotted in the pink dashed line). Figure 3.10d shows the absorption spectrum of the helix now embedded in the matrix described above. The interaction leads to a wide anticrossing between peaks B_1 and B_2, on the order of 300 meV. However, since resonance A is also close in energy, it can also couple to the exciton. At the same time, the excitonic layer itself supports a geometric resonance GR, which still manifests as an uncoupled spectral feature at 2.4 eV. As a consequence, the absorption spectrum is now characterized by five resonances, each with different chirality properties. Tracing which of the split branches corresponds to which original resonance is straightforward, as the hybrid modes still maintain the preference in handedness of the uncoupled ones. In this particular example, the resonances with higher amplitude under RCP incidence (A_1, A_2) originate from resonance A, while higher LCP intensity (B_1, B_2) corresponds to resonance B. Since resonance GR is entirely related to the matrix, it has no handedness and its CD signal is practically zero, as one can see in Fig. 3.10f.

By allowing $\omega_{\rm exc}$ to vary, one can scan the entire energy window of interest, and obtain the CD colour map of Fig. 3.10g. Here, it is more evident how the one band of strong CD shown in Fig. 3.10e has transformed into two, separated by an anticrossing with practically zero CD, following the linear dispersion corresponding to the exciton-polariton resonance of the matrix itself. It is important to note here, that the size of the NPs calls for a rigorous consideration of nonlocal and finite-size effects. However, invoking the analysis done in Sect. 3.2.1, even though the individual chain modes are expected to blueshift and broaden, the coupling strength remains unaffected (see, for instance, Fig. 3.3) [22, 36].

References

1. Schuller JA, Barnard ES, Cai W, Jun YC, White JS, Brongersma ML (2010) Nat Mater 9:193 (2010). https://doi.org/10.1038/nmat2630
2. Baumberg JJ, Aizpurua J, Mikkelsen MH, Smith DR (2019) Nat Mater 18:668. https://doi.org/10.1038/s41563-019-0290-y
3. Dintinger J, Klein S, Bustos F, Barnes WL, Ebbesen TW (2005) Phys Rev B 71:035424. https://doi.org/10.1103/PhysRevB.71.035424
4. Lidzey DG, Bradley DDC, Skolnick MS, Virgili T, Walker S, Whittaker DM (1998) Nature 395:53. https://doi.org/10.1038/25692
5. Yoshie T, Scherer A, Hendrickson J, Khitrova G, Gibbs HM, Rupper G, Ell C, Shchekin OB, Deppe DG (2004) Nature 432:200. https://doi.org/10.1038/nature03119
6. Zheng D, Zhang S, Deng Q, Kang M, Nordlander P, Xu H (2017) Nano Lett 17:3809. https://doi.org/10.1021/acs.nanolett.7b01176

7. García-Etxarri A, Gómez-Medina R, Froufe-Pérez LS, López C, Chantada L, Scheffold F, Aizpurua J, Nieto-Vesperinas M, Sáenz JJ (2011) Opt Express 19:4815. https://doi.org/10.1364/OE.19.004815
8. Tserkezis C, Gonçalves PAD, Wolff C, Todisco F, Busch K, Mortensen NA (2018) Phys Rev B 98:155439. https://doi.org/10.1103/PhysRevB.98.155439
9. Wang H, Ke Y, Xu N, Zhan R, Zheng Z, Wen J, Yan J, Liu P, Chen J, She J, Zhang Y, Liu F, Chen H, Deng S (2016) Nano Lett 16:6886. https://doi.org/10.1021/acs.nanolett.6b02759
10. Todisco F, Malureanu R, Wolff C, Gonçalves PAD, Roberts AS, Mortensen NA, Tserkezis C (2020) Nanophotonics 9:803. https://doi.org/10.1515/nanoph-2019-0444
11. Wang S, Raziman TV, Murai S, Castellanos GW, Bai P, Berghuis AM, Godiksen RH, Curto AG, Gómez Rivas J (2020) J Phys Chem C 124:19196. https://doi.org/10.1021/acs.jpcc.0c02592
12. Baranov DG, Munkhbat B, Länk NO, Verre R, Käll M, Shegai T (2010) Nanophotonics 9:283. https://doi.org/10.1515/nanoph-2019-0372
13. García-Etxarri A, Dionne JA (2013) Phys Rev B 87:235409. https://doi.org/10.1103/PhysRevB.87.235409
14. Baranov DG, Schäfer C, Gorkunov MV (2023) ACS Photon 10:2440. https://doi.org/10.1021/acsphotonics.2c02011
15. Bohren CF, Huffman DR (1983) Absorption and scattering of light by small particles. Wiley
16. Hohenester U, Trügler A (2012) Comput Phys Commun 183:370. https://doi.org/10.1016/j.cpc.2011.09.009
17. Fofang NT, Grady NK, Fan Z, Govorov AO, Halas NJ (2011) Nano Lett 11:1556. https://doi.org/10.1021/nl104352j
18. Antosiewicz TJ, Apell SP, Shegai T (2014) ACS Photon 1:454. https://doi.org/10.1021/ph500032d
19. Raza S, Bozhevolnyi SI, Wubs M, Mortensen NA (2015) J Phys Condens Matter 27:183204. https://doi.org/10.1088/0953-8984/27/18/183204
20. Ritchie R, Marusak A (1966) Surf Sci 4:234. https://doi.org/10.1016/0039-6028(66)90003-3
21. Ford G, Weber W (1984) Phys Rep 113:195. https://doi.org/10.1016/0370-1573(84)90098-X
22. Tserkezis C, Wubs M, Mortensen NA (2018) ACS Photon 5:133. https://doi.org/10.1021/acsphotonics.7b00538
23. Gonçalves PAD, Christensen T, Rivera N, Jauho A-P, Mortensen NA, Soljačić M (2020) Nat Commun 11:366. https://doi.org/10.1038/s41467-019-13820-z
24. Castellanos GW, Murai S, Raziman TV, Wang S, Ramezani M, Curto AG, Gómez Rivas J (2020) ACS Photon 7:1226. https://doi.org/10.1021/acsphotonics.0c00063
25. Green MA (2008) Sol Energy Mater Sol Cells 92:1305. https://doi.org/10.1016/j.solmat.2008.06.009
26. Warning LA, Miandashti AR, McCarthy LA, Zhang Q, Landes CF, Link S (2021) ACS Nano 15:15538. https://doi.org/10.1021/acsnano.1c04992
27. García-Etxarri A, Dionne JA (2013) Phys Rev B 87:235409. https://doi.org/10.1103/PhysRevB.87.235409
28. Lasa-Alonso J, Abujetas DR, Nodar Á, Dionne JA, Sáenz JJ, Molina-Terriza G, Aizpurua J, García-Etxarri A (2020) ACS Photon 7:2978. https://doi.org/10.1021/acsphotonics.0c00611
29. Graf F, Feis J, Garcia-Santiago X, Wegener M, Rockstuhl C, Fernandez-Corbaton I (2019) ACS Photon 6:482. https://doi.org/10.1021/acsphotonics.8b01454
30. Kuzyk A, Schreiber R, Fan Z, Pardatscher G, Roller E-M, Högele A, Simmel FC, Govorov AO, Liedl T (2012) Nature 483:311. https://doi.org/10.1038/nature10889
31. Fernandez-Corbaton I, Zambrana-Puyalto X, Tischler N, Vidal X, Juan ML, Molina-Terriza G (2013) Phys Rev Lett 111:060401. https://doi.org/10.1103/PhysRevLett.111.060401
32. Vázquez-Guardado A, Chanda D (2018) Phys Rev Lett 120:137601. https://doi.org/10.1103/PhysRevLett.120.137601
33. Johnson PB, Christy RW (1972) Phys Rev B 6:4370. https://doi.org/10.1103/PhysRevB.6.4370
34. Liu N, Liedl T (2018) Chem Rev 118:3032. https://doi.org/10.1021/acs.chemrev.7b00225
35. Tserkezis C, Taylor RW, Beitner J, Esteban R, Baumberg JJ, Aizpurua J (2014) Part Part Syst Charact 31:152. https://doi.org/10.1002/ppsc.201300287
36. Tserkezis C, Yeşilyurt ATM, Huang J-S, Mortensen NA (2018) ACS Photon 5:5017. https://doi.org/10.1021/acsphotonics.8b01261

Chapter 4
Weak Coupling of Nanoparticles with Quantum Emitters

Abstract This chapter presents a study of the weak interaction between a single spherical metallic or dielectric scatterer and a quantum emitter. We explore the possibilities of controlling the excitation and emission properties of the emitter based on its position and orientation relative to the structure, highlighting the different capabilities observed when comparing metal and dielectric nanocavities. Additionally, we examine the interaction between quantum emitters and plasmonic structures of a few nanometers under the scope of mesoscopic plasmonics and analyze the emerging nonclassical effects.

The interaction of light with QEs encountered in atoms, molecules or semiconductors lies at the heart of nano-optics. There exist several confined entities that can be visualized as two-level systems, such as organic dye molecules, semiconductor quantum dots, or color centers[1] in wide-bandgap semiconductors. Controlling the radiative properties of QEs plays a significant role, for instance for sensing and imaging in biomedical studies [1–3], as well as in the design of antennas for nanophotonic applications [4, 5]. Ever since Purcell's realization that spontaneous emission is modified by changes in the emitter environment [6], considerable effort has been devoted to fluorescence enhancement via coupling to nearby resonant structures.

So far, we have described how metal NPs can strongly interact with light, and achieve strong field enhancement and confinement of the incident EM field. Due to these characteristics, the concept of plasmon-enhanced fluorescence has remained a common technique for controlling the emission of molecules for decades [7–9]. Several theoretical and experimental works have addressed the control of molecular emission in the vicinity of metallic surfaces employing planar interfaces [10, 11], single NPs [12–14], dimers [15, 16], and other complex geometries [17–20].

Nevertheless, despite the tremendous enhancement of the excitation rate due to the intense near fields around plasmonic NPs, Ohmic losses that lead to poor radiative properties can still be a hurdle, especially when quantum effects in the material

---

[1] Color centers are point defects in the crystal lattice of solids, formed by foreign atoms or vacancies. Due to the existence of allowed transitions, the trapped electrons absorb photons, usually in the visible spectrum, and release them back to the environment through spontaneous emission.

P. E. Stamatopoulou, *Strong Light-Matter Interactions in Extreme Plasmonic and Mie-Resonant Systems*, Springer Theses, https://doi.org/10.1007/978-3-031-67907-0_4

response become important [21–23]. An alternative route is once again sought out in dielectric resonators, either in the context of hybrid plasmonic–dielectric structures that combine the light confinement of plasmonics with the high quality factors of microscale dielectric cavities [24, 25], or in all-dielectric nanostructures [26–28].

In this chapter we present the study published in Publication E, concerning the weak interaction of single spherical metallic and dielectric NPs with QEs. We explore the possibilities of controlling excitation and emission properties depending on the position and orientation of the latter, and discuss the differences between metals and dielectrics. Finally, we investigate the interaction between nanoscale plasmonic NPs and QEs, by including the necessary corrections in our theory, and explore how the observables are affected as a result.

4.1 Mie Theory with Electric-Dipole Excitation

Consider a system consisting of a spherical NP in close proximity to a QE, as depicted in Fig. 4.1a. The QE is excited by a plane wave polarized along the z axis, resulting in the absorption of a photon of energy $\hbar\omega_{\mathrm{exc}}$, and a subsequent decay by emitting a photon of energy $\hbar\omega_{\mathrm{em}} \approx \hbar\omega_{\mathrm{exc}}$, assuming that the QE is well described by a two-level system. To explore the excitation and emission properties of the QE in the structured environment, we begin by revisiting the expressions presented in Sect. 2.3.4.2. By expanding all fields in the basis of spherical waves, we can apply Mie theory and derive analytic expressions for the excitation enhancement η_{exc} and photonic enhancement P/P_0 of the system, for the various positions and orientations of the dipole. Here, we consider the two cases of QE orientation depicted in Fig. 4.1b; parallel orientation $\mathbf{p}_{\mathrm{d}} \parallel \mathbf{r}_{\mathrm{d}}$ (case $\parallel$), and perpendicular orientation $\mathbf{p}_{\mathrm{d}} \perp \mathbf{r}_{\mathrm{d}}$ (case $\perp$) with respect to the position vector. We examine the excitation and emission of the system for 4 different positions of the point dipole: $\mathbf{r}_{\mathrm{d}} = r_{\mathrm{d}}\hat{\mathbf{z}}$ ($\mathrm{A}_{1/2}$), $\mathbf{r}_{\mathrm{d}} = r_{\mathrm{d}}\hat{\mathbf{y}}$ (B), $\mathbf{r}_{\mathrm{d}} = r_{\mathrm{d}}\hat{\mathbf{x}}$ (C), and $\mathbf{r}_{\mathrm{d}} = -r_{\mathrm{d}}\hat{\mathbf{x}}$ (D). In all cases we assume an incident wave of amplitude $\mathbf{E}_0 = E_0\hat{\mathbf{z}}$ propagating along the $\hat{\mathbf{x}}$ axis.

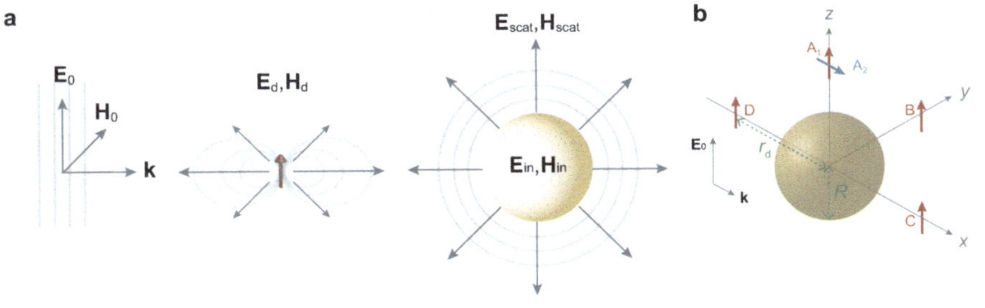

Fig. 4.1 **Weak interaction between a NP and a QE.** **a** Schematic illustration of a system consisting of a spherical NP of radius R and a QE, modeled as a point electric dipole, under plane wave illumination. **b** The different positions $\mathbf{r}_{\mathrm{d}}$ and orientations of the electric dipole moment $\mathbf{p}_{\mathrm{d}}$ discussed in the text

Excitation Enhancement The excitation enhancement of a dipole illuminated by a plane wave of amplitude E_0 in the presence of the spherical NP is found by Eq. (2.108)

$$\eta_{\rm exc} = \frac{|\mathbf{p}_{\rm d} \cdot \mathbf{E}_{\rm tot}(\mathbf{r}_{\rm d})|^2}{|\mathbf{p}_{\rm d} \cdot \mathbf{E}_0(\mathbf{r}_{\rm d})|^2}\,, \tag{4.1}$$

where the total field $\mathbf{E}_{\rm tot}$ includes both $\mathbf{E}_0$, as well as the field scattered off the NP surface. We already have at hand analytic expressions for the incoming and scattered field in spherical-wave expansions, i.e. Eqs. (2.75), that make up the total field at the position of the emitter, which is responsible for its excitation. In all configurations, apart from A_2, the denominator of Eq. (4.1) is equal to $(p_{\rm d}E_0)^2$. In configuration A_2, the denominator is equal to 0, and the electric dipole is not excited by the incident illumination, but only by the fields scattered off the surface of the NP. In this case, we define the excitation enhancement as the numerator of Eq. (4.1) normalized to $(p_{\rm d}E_0)^2$. Taking the above into account, and by inserting Eqs. (2.75) in Eq. (4.1) for the various positions and orientations shown in Fig. 4.1b, we find (see Appendix E.1)

$$\eta_{\rm exc}^{\rm A_1} = \pi \left| \sum_\ell \frac{i^{\ell+1}\sqrt{\ell(\ell+1)(2\ell+1)}}{k_2 r_{\rm d}} \Big[j_\ell(k_2 r_{\rm d}) + T_{{\rm E}\ell} h_\ell^+(k_2 r_{\rm d}) \Big] \mathscr{Y}_{\ell 1}^- \right|^2, \tag{4.2a}$$

$$\begin{aligned}\eta_{\rm exc}^{\rm A_2} = \frac{\pi}{4} \Bigg| \sum_\ell & \frac{i^{\ell+1}\sqrt{2\ell+1}}{k_2 r_{\rm d}} \Big[\Psi_\ell'(k_2 r_{\rm d}) + T_{{\rm E}\ell}\xi_\ell'(k_2 r_{\rm d}) \Big] \left[\mathscr{Y}_{\ell 0}^+ - \sqrt{\frac{(\ell-1)(\ell+2)}{\ell(\ell+1)}}\, \mathscr{Y}_{\ell 2}^+ \right] \\ & - 2 i^\ell \sqrt{\frac{2\ell+1}{\ell(\ell+1)}} \Big[j_\ell(k_2 r_{\rm d}) + T_{{\rm M}\ell} h_\ell^+(k_2 r_{\rm d}) \Big] \mathscr{Y}_{\ell 1}^- \Bigg|^2, \end{aligned} \tag{4.2b}$$

$$\begin{aligned}\eta_{\rm exc}^{\rm B} = \frac{\pi}{4} \Bigg| \sum_\ell & i^{\ell+1}\sqrt{2\ell+1} \Big[j_\ell(k_2 r_{\rm d}) + T_{{\rm M}\ell} h_\ell^+(k_2 r_{\rm d}) \Big] \left[\mathscr{Y}_{\ell 0}^+ - \sqrt{\frac{(\ell-1)(\ell+2)}{\ell(\ell+1)}}\, \mathscr{Y}_{\ell 2}^+ \right] \\ & + \frac{2 i^\ell}{k_2 r_{\rm d}} \sqrt{\frac{2\ell+1}{\ell(\ell+1)}} \Big[\Psi_\ell'(k_2 r_{\rm d}) + T_{{\rm E}\ell}\xi_\ell'(k_2 r_{\rm d}) \Big] \mathscr{Y}_{\ell 1}^- \Bigg|^2, \end{aligned} \tag{4.2c}$$

and

$$\begin{aligned}\eta_{\rm exc}^{\rm C/D} = \frac{1}{4} \Bigg| \sum_\ell (\pm 1)^\ell (2\ell+1) \Bigg\{ & \frac{i^{\ell+1}}{k_2 r_{\rm d}} \Big[\Psi_\ell'(k_2 r_{\rm d}) + T_{{\rm E}\ell}\xi_\ell'(k_2 r_{\rm d}) \Big] \\ & - i^\ell \Big[j_\ell(k_2 r_{\rm d}) + T_{{\rm E}\ell} h_\ell^+(k_2 r_{\rm d}) \Big] \Bigg\} \Bigg|^2, \end{aligned} \tag{4.2d}$$

where we introduced

$$\mathscr{Y}_{\ell j}^\pm = Y_\ell^{-j}\left(\frac{\pi}{2}, 0\right) \pm Y_\ell^j\left(\frac{\pi}{2}, 0\right). \tag{4.3}$$

Note that Eq. (4.1) can be used for the evaluation of the excitation enhancement for any position $\mathbf{r}_\mathrm{d}$ of the QE. The selection of positions displayed in Fig. 4.1b serves a dual purpose: it is representative enough for the sake of illustration, while also substantially simplifying the final analytic expressions, which become independent of m.

Radiative and Non-radiative Photonic Enhancement To examine the emission properties of the system, we assume that both the NP and the QE are already excited by the plane wave, which we now "turn off", and let the system decay. The problem is once again a Mie-scattering problem, with the internal and scattered fields being identical to the ones derived in Eqs. (2.75), but the incident plane wave is replaced by the EM field of the QE. Expanded in the basis of spherical waves, the field of an oscillating dipole takes the form of incoming spherical waves at points in space $r < r_d$, and outgoing everywhere else, i.e.

$$\mathbf{E}(\mathbf{r}) = \sum_{\ell m} \left\{ \left[d^0_{\mathrm{M};\ell m} j_\ell(k_2 r) \mathbf{X}_{\ell m}(\hat{\mathbf{r}}) + \frac{i}{k_2} d^0_{\mathrm{E};\ell m} \nabla \times j_\ell(k_2 r) \mathbf{X}_{\ell m}(\hat{\mathbf{r}}) \right] \Theta(r_d - r) \right.$$
$$\left. + \left[d^+_{\mathrm{M};\ell m} h^+_\ell(k_2 r) \mathbf{X}_{\ell m}(\hat{\mathbf{r}}) + \frac{i}{k_2} d^+_{\mathrm{E};\ell m} \nabla \times h^+_\ell(k_2 r) \mathbf{X}_{\ell m}(\hat{\mathbf{r}}) \right] \Theta(r - r_d) \right\}, \quad (4.4)$$

where $\Theta(x)$ is the Heaviside step function [29], and the expansion coefficients $d^{0/+}$ can be found by the following equations

$$d^0_{\mathrm{E};\ell m} = \frac{i k_2^3 p_\mathrm{d}}{\varepsilon_2 \varepsilon_0} \mathbf{p}_\mathrm{d} \cdot \frac{i}{k_2} \nabla \times h^+_\ell(k_2 r) \mathbf{X}^*_{\ell m}(k_2 r), \quad (4.5a)$$

$$d^0_{\mathrm{M};\ell m} = \frac{i k_2^3 p_\mathrm{d}}{\varepsilon_2 \varepsilon_0} \mathbf{p}_\mathrm{d} \cdot h^+_\ell(k_2 r) \mathbf{X}^*_{\ell m}(k_2 r), \quad (4.5b)$$

$$d^+_{\mathrm{E};\ell m} = \frac{i k_2^3 p_\mathrm{d}}{\varepsilon_2 \varepsilon_0} \mathbf{p}_\mathrm{d} \cdot \frac{i}{k_2} \nabla \times j_\ell(k_2 r) \mathbf{X}^*_{\ell m}(k_2 r), \quad (4.5c)$$

$$d^+_{\mathrm{M};\ell m} = \frac{i k_2^3 p_\mathrm{d}}{\varepsilon_2 \varepsilon_0} \mathbf{p}_\mathrm{d} \cdot j_\ell(k_2 r) \mathbf{X}^*_{\ell m}(k_2 r). \quad (4.5d)$$

As discussed in Sect. 2.3.4.2, the energy of the QE–NP system dissipates by either radiation or via non-radiative processes, in particular due to absorption in the cavity formed by the NP. For the two different dipole orientations shown in Fig. 2.8, the radiative decay rate can be shown to give

$$\frac{P_\mathrm{r}^{\parallel}}{P_0} = \frac{3}{2(k_2 r_\mathrm{d})^2} \sum_\ell \ell(\ell+1)(2\ell+1) \left| j_\ell(k_2 r_\mathrm{d}) + T_{\mathrm{E}\ell} h^+_\ell(k_2 r_\mathrm{d}) \right|^2, \quad (4.6a)$$

$$\frac{P_\mathrm{r}^{\perp}}{P_0} = \frac{3}{4} \sum_\ell (2\ell+1) \left\{ \frac{1}{(k_2 r_\mathrm{d})^2} \left| \Psi'_\ell(k_2 r_\mathrm{d}) + T_{\mathrm{E}\ell} \xi'_\ell(k_2 r_\mathrm{d}) \right|^2 \right.$$
$$\left. + \left| j_\ell(k_2 r_\mathrm{d}) + T_{\mathrm{M}\ell} h^+_\ell(k_2 r_\mathrm{d}) \right|^2 \right\}, \quad (4.6b)$$

while the non-radiative yields

$$\frac{P_{\mathrm{nr}}^{\parallel}}{P_0} = -\frac{3}{2(k_2 r_{\mathrm{d}})^2} \sum_{\ell} \ell(\ell+1)(2\ell+1)\Big[\mathrm{Re}\big\{T_{\mathrm{E}\ell}\big\} + \big|T_{\mathrm{E}\ell}\big|^2\Big]\big|h_{\ell}^{+}(k_2 r_{\mathrm{d}})\big|^2 , \tag{4.7a}$$

$$\frac{P_{\mathrm{nr}}^{\perp}}{P_0} = -\frac{3}{4} \sum_{\ell} (2\ell+1)\Bigg\{\Big[\mathrm{Re}\big\{T_{\mathrm{E}\ell}\big\} + \big|T_{\mathrm{E}\ell}\big|^2\Big]\left|\frac{\xi_{\ell}'(k_2 r_{\mathrm{d}})}{k_2 r_{\mathrm{d}}}\right|^2 + \Big[\mathrm{Re}\big\{T_{\mathrm{M}\ell}\big\} + \big|T_{\mathrm{M}\ell}\big|^2\Big]\big|h_{\ell}^{+}(k_2 r_{\mathrm{d}})\big|^2\Bigg\} . \tag{4.7b}$$

More details regarding the derivation can be found in Appendix E.2.

4.2 Weak Coupling of Metallic Nanoparticles with Quantum Emitters

We consider a silver NP of radius $R = 50$ nm, described by the experimental bulk permittivity of Ref. [30], in close proximity to a QE placed at various positions and orientations, as depicted in Fig. 4.2f. The NP is large enough to serve as a good antenna, as one can deduce from the relative contributions of scattering and absorption to the extinction spectrum in Fig. 4.2a. In the same panel, we observe a strongly radiative ED mode at $\hbar\omega_{\mathrm{ED}} = 3.15$ eV, and a non-radiative EQ mode at $\hbar\omega_{\mathrm{EQ}} = 3.52$ eV. Therefore, we expect the highest fluorescence enhancement at energies close to the ED mode. Indeed, in Fig. 4.2d we observe considerable excitation rate enhancement (dashed curves) for configuration A_1 at this energy, but significantly lower for any other. This is somewhat expected since, in the excitation process, the surface charges are polarized parallel to the incident plane wave polarization, along the z axis, and therefore the near-field exhibits maximum enhancement close to the position of the A_1 and A_2 configurations. However, configuration A_2 is polarized perpendicularly to the incident electric field, so it can only be excited by the scattered field, rather than the incident plane wave, leading to a very low excitation efficiency.

Panel d shows that we obtain the highest value of the fluorescence rate enhancement for configuration A_1. In fact, in any other case the emission process is suppressed rather than boosted. The enhancement of the fluorescence process in configuration A_1 is the result of not only the more effective excitation of the QE, but also of the fact that the NP radiates more efficiently, as can be verified by the radiation patterns of Fig. 4.2e. This claim is further corroborated in Fig. 4.2c, which shows the highest radiative decay rate (solid dark blue curve), leading to a higher quantum yield in Fig. 4.2b compared to any other position and orientation of the QE. In panel b we also observe that the quantum yield (pink curves) does not assume its highest value exactly at the energy of the ED mode, but at a slightly lower energy, due to a higher contrast between absorption and scattering; interestingly, at this energy the absorption cross section in Fig. 4.2a exhibits a small dip.

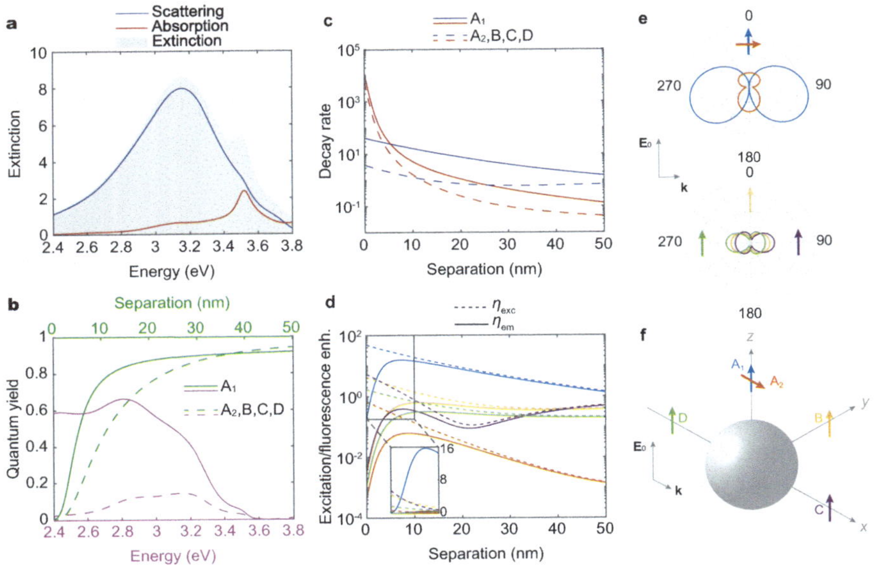

Fig. 4.2 Weak interaction between an Ag NP and a QE. a Normalized extinction (gray shaded area), absorption (dark red curve) and scattering (dark blue curve) cross sections of an Ag sphere of radius $R = 50$ nm, under plane-wave illumination. **b** Quantum yield of the QE for the various configurations shown in panel **f**, plotted versus the separation from the NP surface (green curve) considering energy $\hbar\omega = 3.15$ eV, and versus the energy of the incident wave (magenta curve) considering separation 5 nm. **c** Radiative (dark blue curves) and non-radiative (dark red curves) decay rate enhancement versus NP–QE separation for the various configurations shown in panel **f**, at energy $\hbar\omega = 3.15$ eV. **d** Excitation (dashed curves) and fluorescence enhancement (solid curves) versus NP–QE separation following the same color coding with the configurations shown in panel **f**, at energy $\hbar\omega = 3.15$ eV. The inset shows the enhancement factors in linear scale for separations 0–10 nm. **e** Far-field radiation patterns of the scattered field versus the polar angle θ considering an energy of $\hbar\omega = 3.15$ eV and separation 5 nm, for the QE configurations shown in panel **f**. **f** Schematic illustration of a system consisting of the spherical Ag NP and the QE, modeled as a point electric dipole at various positions and orientations (A_1, A_2, B, C, D), excited by a plane wave polarized along the z axis, propagating in the x direction

In Fig. 4.2d we observe that, as the NP–QE separation is decreased, the excitation enhancement η_{exc} increases monotonically. The fluorescence enhancement η_{em} follows a similar trend at larger distances, until it reaches a maximum of $\eta_{em} \approx 16$ (configuration A_1) at separation 7.3 nm, and drops for further decrease of the separation. Recalling Eq. (2.116): $\eta_{em} = \eta_{exc}\,\mathcal{Q}/\mathcal{Q}_0$, this decrease is explained by the drop, in Fig. 4.2b (green curves), of the quantum yield from a high almost constant value of 0.8 for separations above 10 nm to 0 as the QE approaches the surface of the NP. At such small separations the QE accesses the near field of the NP, and can therefore excite the non-radiative higher-order multipoles [31]. This is confirmed in Fig. 4.2c, where for separations below 5 nm we observe a steep increase in the non-radiative decay rate of the QE (solid dark red curve), meaning that it decays mostly due to energy absorption from the NP. Therefore, there is an optimal distance between the QE and the NP: small enough so that the QE is excited more efficiently from the strong scattered fields of the NP, while large enough so that the dark high-order plasmonic modes are not excited.

4.2.1 Nonlocal Effects in Plasmon-Enhanced Fluorescence

Let us now explore how the excitation and fluorescence enhancement is modified in NPs of nanometric sizes, when nonlocal effects become important. As a testbed we consider a Na NP of radius $R = 5$ nm, whose permittivity is given by the Drude model of Eq. (2.54) for $\hbar\omega_p = 5.89$ eV, $\hbar\gamma_p = 100$ meV. We treat the nonlocal response of the material employing the Feibelman SRF, using the d-parameters provided in Ref. [32] (see Fig. 2.10c). Due to the spill-out effect, the plasmonic mode predicted at 3.4 eV in the LRA calculation (dashed curve) in Fig. 4.3a redshifts to 3.3 eV in the SRF approach (solid curve). A NP of this size is not a good antenna, and it mostly absorbs the incident fields (see for instance Fig. 2.5d), leading to very low quantum yields. However, the smaller the NP the higher the field enhancement at its surface, resulting eventually in an appreciable enhancement of the fluorescence rate at the resonant energy, as demonstrated in Fig. 4.3b for the three QE configurations shown in the inset. In the same panel we see that the quantum-corrected Feibelman calculation predicts a significantly lower fluorescence enhancement compared to the LRA model, dropping from a factor of approximately 8 to 1.2 for the higher-efficiency configuration A_I.

Figure 4.3c–f show the excitation and emission properties of the QE versus its separation from the NP surface as reproduced in LRA (dashed curves) and the SRF model (solid curves). Indeed, in Fig. 4.3c, for decreasing separation, we observe increasingly high enhancement factors in the excitation of the QE for configurations A_I and B, but slightly lower in the SRF model compared to the LRA. In panel d we find an exceptional decrease of the optimal fluorescence enhancement from the LRA to the SRF prediction, from 21 to 1.25 in configuration A_I, and from 5 to less than 1 in configuration B, owing to the surface-enhanced Landau damping and the spreading of the charges at the NP boundary. This feature agrees with previous works reporting saturation and, eventually, decline in the field enhancement within nanogaps between metallic structures, with Landau damping setting the ultimate limit of obtainable plasmonic enhancement [33, 34].

According to Fig. 4.3d, the highest fluorescence enhancement is achieved in configuration A_I ($\eta_{em} \approx 1.25$) at separation 1.62 nm, which is larger by ~ 1 nm compared to the LRA prediction (0.54 nm); this can be ascribed to the effective decrease of the NP–QE separation due to the electron-density spill-out. As a result, compared to the LRA calculation, the SRF model predicts a lower non-radiative decay rate in the far-field region and higher in the near-field region, below a separation of 3 nm (compare the cyan/pink solid curves with the dashed ones in Fig. 4.3f). A similar

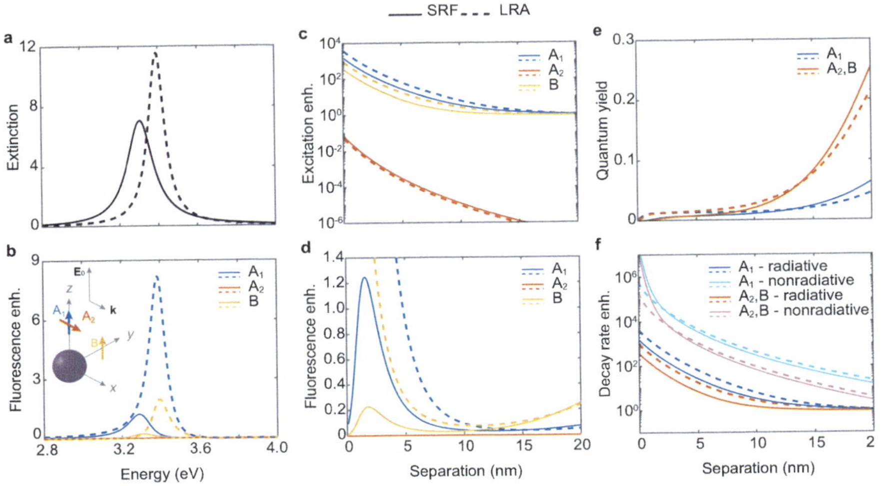

Fig. 4.3 Nonlocal effects in the interaction between a Na NP and a QE. **a** Normalized extinction cross section under plane-wave illumination of a Na NP of radius $R = 5$ nm, and dielectric function given by the Drude model of Eq. (2.54), for $\hbar\omega_p = 5.89$ eV, $\hbar\gamma_p = 100$ meV. The solid curves show results obtained with the Feibelman (SRF) formalism, compared to LRA with the dashed curves. **b** Fluorescence enhancement versus energy for the three QE configurations (A_1, A_2, B) shown in the inset, placed at distance 5 nm from the NP surface. The system is excited by a plane wave polarized along the z axis and propagating in the x direction. **c** Excitation enhancement versus NP–QE separation for the various configurations of the QE, following the same color coding as the inset of panel b. **d** Fluorescence enhancement versus NP–QE separation for the various QE configurations. **e** Quantum yield versus NP–QE separation for the various QE configurations. **f** Radiative (blue and cyan curves) and non-radiative (orange and pink curves) decay rate enhancement versus NP–QE separation for the various QE configurations. In all panels the solid lines correspond to the results produced using the SRF model and the d-parameters provided in Ref. [32] (see Fig. 2.10c), whereas dashed lines show the LRA predictions. In panels c–f we calculate the relevant quantities at the resonant energy predicted by each model, i.e. at 3.4 eV in LRA and at 3.3 eV in the SRF calculation

trend can be observed in panel e, where a larger value for the quantum yield is predicted in SRF at larger separations, while smaller values are obtained for decreasing separation, as compared to the LRA results. Interestingly, in contrast to the larger plasmonic NP in Fig. 4.2b, here we achieve a higher quantum yield in the tangential configurations A_2 and B rather than the radial A_1 in both the LRA and the SRF calculation, meaning that the process of absorption from the NP is more efficient in the latter orientation.

It is important to note here, that the prediction of the SRF model is unreliable for extremely small separations. As discussed in Sect. 2.3.5.5, the validity of the Feibelman formalism is restricted to the long-wavelength approximation, i.e. as long as $R/\ell \gg d_\perp \sim$ Å. As the QE approaches the NP and modes of higher order and smaller wavelength are excited, this condition is compromised. Besides, placing the QE inside the evanescent electron-density tail goes beyond CED and requires a purely quantum description. In Publication D we demonstrate that for Na NPs of radius $R = 3.5$ nm the Feibelman model fails for separations between the NP and the QE of less than 1.5 nm, when compared to TDDFT calculations. Given that the NP studied in Fig. 4.3 is larger, we can deduce that the position of optimal fluorescence enhancement at 1.62 nm is an accurate estimate.

4.3 Weak Coupling of Dielectric Nanoparticles with Quantum Emitters

For the control of the emission properties of QEs, dielectric NPs are oftentimes chosen over plasmonic, owing to their ability to support low-loss electric and magnetic Mie resonances of multiple radial orders. Modulating the overlap of said modes enables the steering and switching of light, a property that is highly desirable for the design of nanoantennas. In fact, the developing area of all-dielectric nanoantennas has flourished over the last two decades [35, 36], highlighting high-index dielectrics as suitable platforms for a wide range of applications besides fluorescence enhancement [28], such as for high-harmonic generation [37] and nonclassical light emission [38].

Figure 4.4 explores the excitation and fluorescence enhancement of QEs in close proximity to a high-refractive index NP—here a silicon nanosphere with radius $R = 85$ nm—described by the experimental permittivity of Ref. [39]. The extinction cross section presented in Fig. 4.4a is decomposed into its scattering (dark blue curve) and absorption contributions (dark red curve), showcasing that mesoscopic silicon NPs mostly scatter the incident light, and they are indeed excellent antennas. In particular, at the energy of the MD mode, at which we expect significant field enhancement, ($\hbar\omega = 1.84$ eV) the difference between the scattering and absorption cross sections is very large, implying a high quantum yield. This is verified in panel b, where we plot the quantum yield for the various QE configurations illustrated in panel f, showing a very high—nearly unity—quantum yield for separations larger than 10 nm at this energy. Interestingly, in the same panel we observe an almost constant high quantum yield value for a fixed position of the QE in configuration A_1, that drops at the energy where the dark MQ mode is excited. In contrast, in all other configurations the quantum yield features peaks close to the energies at which the radiative MD and ED modes are excited.

Given the high quantum yield of the MD mode for all QE positions and orientations, in Fig. 4.4d we calculate the enhancement factors of the excitation and fluorescence processes at the energy of the MD resonance. As expected, the excitation rate enhancement (dashed curves) increases monotonically as the NP–QE separation is reduced in all configurations, apart from B. The reason is reflected in the top panel in Fig. 2.6e; the scattered electric field originates from a displacement current loop, circulating inside the NP, that has leaked to the outside, and thus maintains

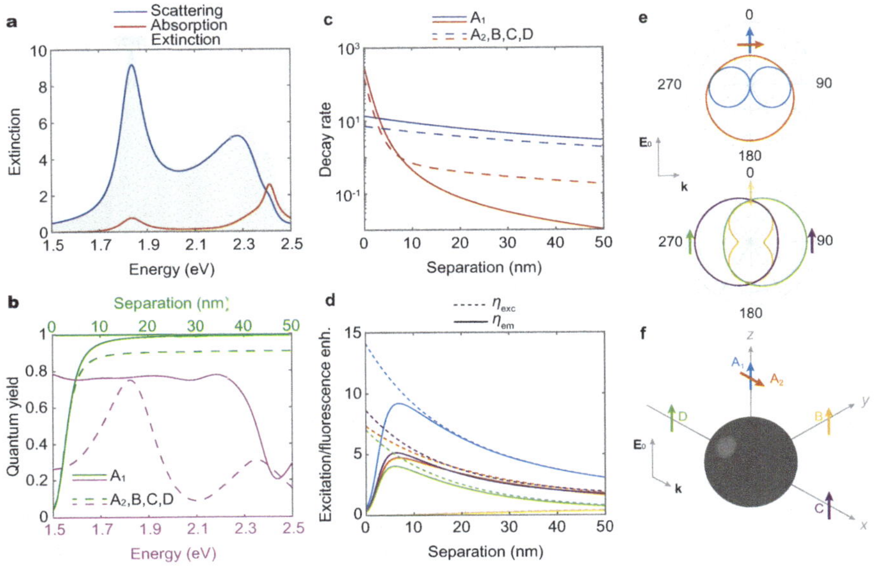

Fig. 4.4 **Weak interaction between a Si NP and a QE.** **a** Normalized extinction (gray shaded area), absorption (dark red curve) and scattering (dark blue curve) cross sections of a Si sphere of radius $R = 85$ nm, under plane-wave illumination. **b** Quantum yield of the QE for the various configurations shown in panel **f**, plotted versus the separation from the NP surface (green curve) considering energy $\hbar\omega = 1.84$ eV, and versus the energy of the incident wave (magenta curve) considering separation 5 nm. **c** Radiative (dark blue curves) and non-radiative (dark red curves) decay rate enhancement versus NP–QE separation for the various QE configurations, at energy $\hbar\omega = 1.84$ eV. **d** Excitation (dashed curves) and fluorescence enhancement (solid curves) versus NP–QE separation, at energy $\hbar\omega = 1.84$ eV. The curves follow the same color coding as the QE configurations shown in panel **f**. **e** Far-field radiation patterns of the scattered field versus the polar angle θ considering energy $\hbar\omega = 1.84$ eV and separation 5 nm, for the various QE configurations. **f** Schematic illustration of a system consisting of the spherical Si NP and the QE, modeled as a point electric dipole at various positions and orientations (A_1, A_2, B, C, D), excited by a plane wave polarized along the z axis, propagating in the x direction

the circulation. It is, therefore, enhanced predominantly along the xz plane. Moreover, Fig. 4.4d reveals that the highest fluorescence rate enhancement (solid curves) is obtained for configuration A_1, at separation of approximately 7 nm. Below this value, an increasing number of non-radiative higher-order modes is excited, which is also corroborated by the decay rates plotted in panel c; for separations below 10 nm the non-radiative decay rate increases exponentially, until it surpasses the radiative decay rate at around 3 nm. Overall the fluorescence enhancement factors in panel d do not exceed a value of ~ 10 for any configuration, which is less than the equivalent result in the plasmonic NP of Sect. 4.2 ($\eta_{em} \sim 16$), and even smaller than the factors of the order of 200 which can be easily achieved with metal-dielectric core–shell

NPs [21]. The reason is that the field enhancement and confinement achieved by dielectric NPs is typically lower compared to their plasmonic counterparts.

In conclusion, while dielectric NPs cannot compete with plasmonic ones in terms of the absolute enhancement achievable, they are more flexible with regards to QE positioning and orientation. They feature great versatility, since, depending on the frequency of the incident light, they can provide either omnidirectional or unidirectional emission, which provides a degree of controlability and selectivity in enhancing or suppressing the emission of nearby emitters. Directional scattering is also possible with plasmonic nanoantennas, nonetheless it requires the integration of quite sophisticated designs [19].

References

1. Lakowicz JR (2006) Plasmonics 1:5. https://doi.org/10.1007/s11468-005-9002-3
2. Deng W, Xie F, Baltar HTMCM, Goldys EM (2013) Phys Chem Chem Phys 15:15695. https://doi.org/10.1039/C3CP50206F
3. Darvill D, Centeno A, Xie F (2013) Phys Chem Chem Phys 15:15709. https://doi.org/10.1039/C3CP50415H
4. Giannini V, Fernández-Domínguez AI, Heck SC, Maier SA (2011) Chem Rev 111:3888. https://doi.org/10.1021/cr1002672
5. Bharadwaj P, Deutsch B, Novotny L (2009) Adv Opt Photon 1:438. https://doi.org/10.1364/AOP.1.000438
6. Purcell EM, Torrey HC, Pound RV (1946) Phys Rev 69:37. https://doi.org/10.1103/PhysRev.69.37
7. Ford G, Weber W (1984) Phys Rep 113:195. https://doi.org/10.1016/0370-1573(84)90098-X
8. Drexhage K (1970) Lumin J 1–2:693. https://doi.org/10.1016/0022-2313(70)90082-7
9. Chance RR, Prock A, Silbey R (1978) Molecular fluorescence and energy transfer near interfaces. In: Advances in chemical physics. Wiley, pp 1–65
10. Amos RM, Barnes WL (1997) Phys Rev B 55:7249. https://doi.org/10.1103/PhysRevB.55.7249
11. Lakowicz JR, Malicka J, Gryczynski I, Gryczynski Z (2003) Biochem BioPhys Res Commun 307:435. https://doi.org/10.1016/S0006-291X(03)01214-2
12. Ruppin R (1982) J Chem Phys 76:1681. https://doi.org/10.1063/1.443196
13. Kühn S, Håkanson U, Rogobete L, Sandoghdar V (2006) Phys Rev Lett 97:017402. https://doi.org/10.1103/PhysRevLett.97.017402
14. Taminiau T, Stefani F, Segerink FB, Van Hulst N (2008) Nat Photon 2:234. https://doi.org/10.1038/nphoton.2008.32
15. Liaw J-W, Chen J-H, Chen C-S, Kuo M-K (2009) Opt Express 17:13532. https://doi.org/10.1364/OE.17.013532
16. Kinkhabwala A, Yu Z, Fan S, Avlasevich Y, Müllen K, Moerner W (2009) Nat Photon 3:654. https://doi.org/10.1038/nphoton.2009.187
17. Blanco LA, García de Abajo FJ (2004) Phys Rev B 69:205414. https://doi.org/10.1103/PhysRevB.69.205414
18. Acuna G, Möller F, Holzmeister P, Beater S, Lalkens B, Tinnefeld P (2012) Science 338:506. https://doi.org/10.1126/science.1228638
19. Curto AG, Volpe G, Taminiau TH, Kreuzer MP, Quidant R, Van Hulst NF (2010) Science 329:930. https://doi.org/10.1126/science.1191922
20. Beermann J, Novikov SM, Albrektsen O, Nielsen MG, Bozhevolnyi SI (2009) J Opt Soc Am B 26:2370. https://doi.org/10.1364/JOSAB.26.002370

21. Tserkezis C, Stefanou N, Wubs M, Mortensen NA (2016) Nanoscale 8:17532. https://doi.org/10.1039/C6NR06393D
22. Tserkezis C, Mortensen NA, Wubs M (2017) Phys Rev B 96:085413. https://doi.org/10.1103/PhysRevB.96.085413
23. Jurga R, D'Agostino S, Della Sala F, Ciracì C (2017) J Phys Chem C 121:22361. https://doi.org/10.1021/acs.jpcc.7b07462
24. Yang Y, Miller OD, Christensen T, Joannopoulos JD, Soljačić M (2017) Nano Lett 17:3238. https://doi.org/10.1021/acs.nanolett.7b00852
25. Palstra IM, Doeleman HM, Koenderink AF (2019) Nanophotonics 8:1513. https://doi.org/10.1515/nanoph-2019-0062
26. Albella P, Poyli MA, Schmidt MK, Maier SA, Moreno F, Sáenz JJ, Aizpurua J (2013) J Phys Chem C 117:13573. https://doi.org/10.1021/jp4027018
27. Krasnok A, Glybovski S, Petrov M, Makarov S, Savelev R, Belov P, Simovski C, Kivshar Y (2016) Appl Phys Lett 108:211105. https://doi.org/10.1063/1.4952740
28. Regmi R, Berthelot J, Winkler PM, Mivelle M, Proust J, Bedu F, Ozerov I, Begou T, Lumeau J, Rigneault H, García-Parajó MF, Bidault S, Wenger J, Bonod N (2016) Nano Lett 16:5143. https://doi.org/10.1021/acs.nanolett.6b02076
29. Olver FWJ, Daalhuis ABO, Lozier DW, Schneider BI, Boisvert RF, Clark CW, Mille BVSBR, Cohl HS, McClain MA (2023) NIST Digital library of mathematical functions. Release 1.1.9 of 2023-03-15
30. Johnson PB, Christy RW (1972) Phys Rev B 6:4370. https://doi.org/10.1103/PhysRevB.6.4370
31. Delga A, Feist J, Bravo-Abad J, Garcia-Vidal FJ (2014) Phys Rev Lett 112:253601. https://doi.org/10.1103/PhysRevLett.112.253601
32. Gonçalves PAD, Christensen T, Rivera N, Jauho A-P, Mortensen NA, Soljačić M (2020) Nat Commun 11:366. https://doi.org/10.1038/s41467-019-13820-z
33. Ciracì C, Hill RT, Mock JJ, Urzhumov Y, Fernández-Domínguez AI, Maier SA, Pendry JB, Chilkoti A, Smith DR (2012) Science 337:1072. https://doi.org/10.1126/science.1224823
34. Khurgin J, Tsai W-Y, Tsai DP, Sun G (2017) ACS Photon 4:2871. https://doi.org/10.1021/acsphotonics.7b00860
35. Baranov DG, Zuev DA, Lepeshov SI, Kotov OV, Krasnok AE, Evlyukhin AB, Chichkov BN (2017) Optica 4:814. https://doi.org/10.1364/OPTICA.4.000814
36. Staude I, Pertsch T, Kivshar YS (2019) ACS Photon 6:802. https://doi.org/10.1021/acsphotonics.8b01326
37. Sain B, Meier C, Zentgraf T (2019) Adv Photon 1:024002. https://doi.org/10.1117/1.AP.1.2.024002
38. Wang K, Titchener JG, Kruk SS, Xu L, Chung H-P, Parry M, Kravchenko II, Chen Y-H, Solntsev AS, Kivshar YS, Neshev DN, Sukhorukov AA (2018) Science 361:1104. https://doi.org/10.1126/science.aat8196
39. Green MA (2008) Sol Energy Mater Sol Cells 92:1305. https://doi.org/10.1016/j.solmat.2008.06.009

Chapter 5
Interaction of Nanoparticles with Electron Beams

Abstract This chapter presents an analytic method for the study of spherical nanoparticles excited by electron beams. We derive expressions for the energy loss and photon emission probability, applicable for electron beams of relativistic velocities passing by or through any spherical isotropic scatterer. Here we apply the theory to simulate cathodoluminescence (CL) and electron energy-loss spectra (EELS) of metallic and dielectric nanoparticles, and compare the results with numerical methods and experimental data. We discuss the emergence of other sources of energy loss, different to the optical excitations of the structure, including interband transitions, Cherenkov and transition radiation.

Even though electron-beam spectroscopy has been a well-established technique for decades, recent developments in data processing have allowed mapping plasmons with nanometer resolution, thus revitalizing it as a groundbreaking tool for the optical characterization of matter [1, 2]. Fast electrons passing by or through a sample undergo energy loss due to energy transfer to the optical modes sustained in the material [3]. From localized and propagating surface plasmons in metallic structures [4–7] to Mie resonances in dielectric nanocavities [8, 9] and phonon polaritons in polar crystals [10–12], electron beams play an instrumental role in the study of collective excitations across a wide spectral range that spans from ultraviolet to far-infrared frequencies.

Electrons are ideal probes for exploring the optical properties of truly nanoscale structures. With their diffraction limit ultimately determined by the de Broglie wavelength, they perform better than any light-based spectroscopy, achieving atomic spatial resolution and sub-meV energy resolution [13, 14]. In EELS highly energetic electrons (30–300 keV) are directed at the structure under study, while the energy loss is measured in a transmission electron microscope (TEM) setup, enabling the detection of both radiative and non-radiative processes [15]. Unlike far-field measurements, EELS can capture dark modes that do not couple to far-field radiation, including longitudinal BPs [16], breathing modes [17], high order optical modes [18] or antibonding modes in nanoparticle dimers [19].

TEM measurements provide the best combination of space and energy resolution, but require thin samples, so that electrons can pass through with minimal deflection. Optical modes in thicker structures can be probed in CL spectroscopy, performed in

P. E. Stamatopoulou, *Strong Light-Matter Interactions in Extreme Plasmonic and Mie-Resonant Systems*, Springer Theses, https://doi.org/10.1007/978-3-031-67907-0_5

scanning electron microscopes (SEMs) at intermediate beam energies (1–50 keV) [3, 13]. In CL measurements what is collected is the far-field photon emission originating from the radiative decay of the excited modes in the sample, while the light emitted can be further analyzed through angle-and polarization-resolved CL measurements [8, 20]. Recent advances in electron spectroscopy techniques have introduced temporal resolution in EEL and CL spectra, launching the field of ultrafast electron microscopy (UEM) [21, 22].

Considering the remarkable advancement in electron spectroscopy techniques and the capabilities unfolding, there is a clear demand for robust theoretical and computational tools to interpret the plethora of experimental data. In this regard, analytic methods have been devised for simple geometries, modeling the electron as a moving point charge in the presence of a polarizable medium. Initially, the first theories were developed within the nonretarded approximation for describing plasmons on thin films [4], but they were quickly generalized to account for optical excitations in various media and geometries [23, 24], while also considering retardation effects [25, 26]. Quantum approaches [27, 28] and analytic solutions that account for relativistic effects for simple geometries were subsequently developed [29, 30], enabling the combination of swift electron beams of relativistic velocities with both common and less conventional materials, such as dielectric media, polar crystals, graphene and other two-dimensional media [9, 12, 31, 32]. However, most theoretical descriptions have focused on aloof electron trajectories, that do not intersect the sample, in contrast to experimental practices, where the electron beam is typically scanned over the sample area [15]. Yet, aloof trajectories cannot capture mechanisms related to the bulk properties of the material, for instance bulk plasmons and phonons [10, 19], or other sources of electron-induced radiation, like Cherenkov (CR) or transition radiation (TR) [29, 33–35].

In this chapter we introduce an analytic method for the study of spherical NPs excited by aloof or penetrating electron beams. Based on Mie theory, and building upon the work of García de Abajo for non-penetrating electrons [30], we derive analytic expressions for the energy loss and photon emission probability, extending their applicability to encompass penetrating trajectories. The method is general and can be used for analyzing any local and isotropic material. Here we apply the theory to study the optical response of a metallic and a dielectric NP. We provide a comparison between the analytic EEL and CL spectra of a plasmonic nanosphere, with simulated results obtained with the DGTD method, based on the results of Preprint A. In the final section we demonstrate a theoretical analysis, accompanied with experimental measurements, regarding the interplay of TR and Mie resonances in the CL spectra of Si nanospheres, which is the topic of Publication D.

5.1 Mie Theory with Swift Electrons

Consider the system depicted in Fig. 5.1: A spherical NP of radius R suspended in air is traversed by a swift electron beam, that excites the optical modes sustained in the structure and loses energy in the process. We presume that the beam is fast enough

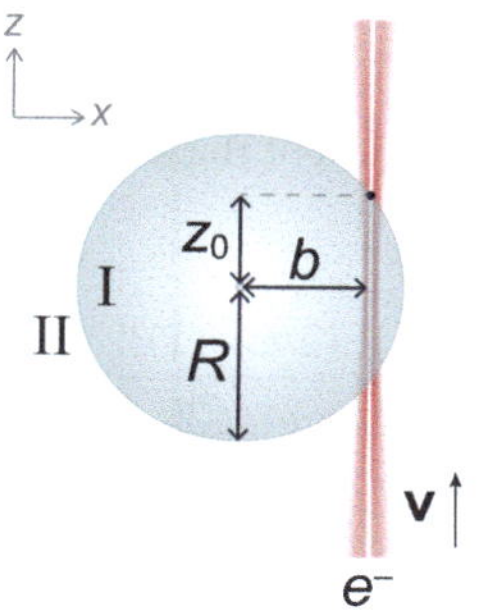

Fig. 5.1 Nanosphere excited by swift electron beam. Schematic illustration of a fast electron traveling with velocity $\mathbf{v} = v\hat{\mathbf{z}}$, through a spherical NP of radius R. We assume that the electron travels on the xz plane, entering the NP at point ($x = b$, $z = -z_0$) and exiting at ($x = b$, $z = z_0$). The region inside the NP, labeled as I, is characterized by optical parameters ε_1 and $\mu_1 = 1$, while label II refers to the surrounding medium (air with $\varepsilon_2 = \mu_2 = 1$)

so that its energy and momentum are much larger compared to those of the optical modes excited in the NP. Therefore, we can invoke the non-recoil approximation, stating that the recoil experienced by the fast electrons in a single scattering event is negligible, and assume that they travel with constant velocity all throughout their motion [36]. We model the electron beam as a single point particle that carries the elementary charge $-e$. Without loss of generality (due to the spherical symmetry of the structure), we assume that it travels along the z-axis, with constant velocity $\mathbf{v} = v\hat{\mathbf{z}}$, following thus a trajectory $\mathbf{r}_e = \mathbf{r}_0 + \mathbf{v}t$, where $\mathbf{r}_0 = (b, \phi = 0, z = -\infty)$ is its initial position in cylindrical coordinates. Here, the impact parameter b indicates the distance between the electron-trajectory axis and the center of the sphere, as shown in the schematics of Fig. 5.1.

In free space, the electric field of the electron, expanded in the basis of spherical waves, can be written as (see Appendix F.1 for a detailed derivation).

Box 5.1 | Electric field of an electron traveling with constant velocity in free space

$$\mathbf{E}_e(\mathbf{r}) = \sum_{\ell m} \left\{ b^0_{\ell m} j_\ell(k_0 r) \mathbf{X}_{\ell m}(\hat{\mathbf{r}}) + \frac{i}{k_0} a^0_{\ell m} \nabla \times j_\ell(k_0 r) \mathbf{X}_{\ell m}(\hat{\mathbf{r}}) \right\}, \tag{5.1}$$

with expansion coefficients

$$b^0_{\ell m} = -\frac{ie}{\varepsilon_0} \frac{k_0^2}{\omega} \frac{m \mathscr{M}^+_{\ell m}}{\sqrt{\ell(\ell+1)}} K_m\left(\frac{\omega b}{v\gamma_0}\right), \tag{5.2a}$$

$$a^0_{\ell m} = \frac{ie}{\varepsilon_0} \frac{k_0^2}{\omega \gamma_0 \beta} \frac{\mathscr{N}^+_{\ell m}}{\sqrt{\ell(\ell+1)}} K_m\left(\frac{\omega b}{v\gamma_0}\right), \tag{5.2b}$$

where $\beta = v/c$ and $\gamma_0 = 1/\sqrt{1-\beta^2}$ are the Lorentz kinematic factors, and K_m is the modified Bessel function of the second kind. In Eq. (5.2a) we have set

$$\mathscr{M}^{+}_{\ell m} = i^{\ell+m} \sqrt{\frac{2\ell+1}{\pi} \frac{(\ell-m)!}{(\ell+m)!} \frac{(2m-1)!!}{(\gamma_0 \beta)^m}} G^{m+1/2}_{\ell-m}\left(\frac{1}{\beta}\right) , \quad \text{for } m \geq 0 , \tag{5.3a}$$

while for $m < 0$, we can make use of the relation

$$\mathscr{M}^{+}_{\ell -m} = (-1)^m \mathscr{M}^{+}_{\ell m} . \tag{5.3b}$$

In Eq. (5.2b) we have set

$$\mathscr{N}^{+}_{\ell m} = \alpha^{m}_{\ell} \mathscr{M}^{+}_{\ell m+1} - \alpha^{-m}_{\ell} \mathscr{M}^{+}_{\ell m-1} , \tag{5.3c}$$

where $\alpha^m_\ell = \sqrt{(\ell-m)(\ell+m+1)}/2$, and $G^{m+1/2}_{\ell-m}$ is the Gegenbauer polynomial [37].

For aloof electron trajectories ($b \geq R$), the field of the moving electron is given by the expressions above. If the electron passes through the NP ($b < R$), the field generated by the electron while in air is, of course, different to the one generated during its motion within the polarizable medium. Assuming that the electron enters the NP at point ($x = b$, $-z_0 = -\sqrt{R^2 - b^2}$) and exits at ($x = b$, z_0) (see Fig. 5.1), the field created by the electron while traveling in air is once again given by Eq. (5.1)

$$\mathbf{E}^{\mathrm{II}}_e(\mathbf{r}) = \sum_{\ell m} \left\{ b^{0,\mathrm{II}}_{\ell m} j_\ell(k_0 r) \mathbf{X}_{\ell m}(\hat{\mathbf{r}}) + \frac{i}{k_0} a^{0,\mathrm{II}}_{\ell m} \nabla \times j_\ell(k_0 r) \mathbf{X}_{\ell m}(\hat{\mathbf{r}}) \right\} , \tag{5.4}$$

but with modified expansion coefficients

$$b^{0,\mathrm{II}}_{\ell m} = -\frac{ie\,k_0^2}{\varepsilon_0\,\omega} \frac{m}{\sqrt{\ell(\ell+1)}} \left[\mathscr{M}^{+}_{\ell m} K_m\left(\frac{\omega b}{v\gamma_0}\right) - i k_0 \int_{-z_0}^{z_0} dz\, e^{i\omega z/v} h^{+}_{\ell}(k_0 r) Y^m_\ell(\theta, 0) \right] , \tag{5.5a}$$

and

$$a^{0,\mathrm{II}}_{\ell m} = \frac{ie\,k_0^2}{\varepsilon_0\,\omega} \frac{1}{\sqrt{\ell(\ell+1)}} \left[\frac{\mathscr{N}^{+}_{\ell m}}{\beta\gamma_0} K_m\left(\frac{\omega b}{v\gamma_0}\right) - \frac{i}{b} \int_{-z_0}^{z_0} dz\, e^{i\omega z/v} \left\{ \mathscr{H}^{+}_{\ell m}(k_0 z) + \mathscr{H}^{-}_{\ell m}(k_0 z) \right\} \right] , \tag{5.5b}$$

where we have set $r = \sqrt{b^2 + z^2}$ and $\theta = \arccos(z/r)$, and function $\mathscr{H}^{\pm}_{\ell m}(k_0 z)$ is given by Eq. (F.76), here for $f_\ell = h^+_\ell$ and $k_i = k_0$. On the other hand, the field created by the electron while traveling inside the NP with wavevector $k_1 = \sqrt{\varepsilon_1} k_0$ is obtained by

$$\mathbf{E}_e^{\mathrm{I}}(\mathbf{r}) = \sum_{\ell m} \left\{ b_{\ell m}^{0,\mathrm{I}} h_\ell^+(k_1 r) \mathbf{X}_{\ell m}(\hat{\mathbf{r}}) + \frac{i}{k_1} a_{\ell m}^{0,\mathrm{I}} \nabla \times h_\ell^+(k_1 r) \mathbf{X}_{\ell m}(\hat{\mathbf{r}}) \right\} , \tag{5.6}$$

with expansion coefficients

$$b_{\ell m}^{0,\mathrm{I}} = -\frac{i e \, k_0^2}{\varepsilon_0 \, \omega} \frac{m}{\sqrt{\ell(\ell+1)}} i k_1 \int_{-z_0}^{z_0} dz \, \mathrm{e}^{i\omega z/v} j_\ell(k_1 r) Y_\ell^m(\theta, 0) , \tag{5.7a}$$

and

$$a_{\ell m}^{0,\mathrm{I}} = \frac{i e \, k_0^2}{\varepsilon_0 \, \omega} \frac{1}{\sqrt{\ell(\ell+1)}} \frac{i}{b} \int_{-z_0}^{z_0} dz \, \mathrm{e}^{i\omega z/v} \left\{ \mathscr{J}_{\ell m}^-(k_1 z) + \mathscr{J}_{\ell m}^+(k_1 z) \right\} . \tag{5.7b}$$

Here, function $\mathscr{J}_{\ell m}^{\pm}(k_1 z)$ is given by Eq. (F.76) for $f_\ell = j_\ell$ and $k_i = k_1$. We note that, by setting $z_0 = 0$ in Eqs. (5.4) and (5.6), and in the coefficients therein, we reproduce the formulas in Box 5.1 which hold for the case of an aloof electron.

The optical modes excited by the electron generate an induced electric field $\mathbf{E}_{\mathrm{ind}}$ and magnetic field $\mathbf{H}_{\mathrm{ind}}$ inside (region I in Fig. 5.1) and outside (region II) the NP, and ultimately release radiation to the environment. As usual, one may decompose the induced EM field into TE and TM components (see Appendix F). In region II, the induced fields take the general form of outgoing spherical waves,

$$\mathbf{E}_{\mathrm{B}}^{\mathrm{II}}(\mathbf{r}) = \sum_{\ell m} \left\{ b_{\ell m}^{\mathrm{II}} h_\ell^+(k_0 r) \mathbf{X}_{\ell m}(\hat{\mathbf{r}}) + \frac{i}{k_0} a_{\ell m}^{\mathrm{II}} \nabla \times h_\ell^+(k_0 r) \mathbf{X}_{\ell m}(\hat{\mathbf{r}}) \right\} , \tag{5.8a}$$

$$\mathbf{H}_{\mathrm{B}}^{\mathrm{II}}(\mathbf{r}) = \frac{1}{Z_0} \sum_{\ell m} \left\{ a_{\ell m}^{\mathrm{II}} h_\ell^+(k_0 r) \mathbf{X}_{\ell m}(\hat{\mathbf{r}}) - \frac{i}{k_0} b_{\ell m}^{\mathrm{II}} \nabla \times h_\ell^+(k_0 r) \mathbf{X}_{\ell m}(\hat{\mathbf{r}}) \right\} , \tag{5.8b}$$

whereas in region I they take the form of standing waves

$$\mathbf{E}_{\mathrm{B}}^{\mathrm{I}}(\mathbf{r}) = \sum_{\ell m} \left\{ b_{\ell m}^{\mathrm{I}} j_\ell(k_1 r) \mathbf{X}_{\ell m}(\hat{\mathbf{r}}) + \frac{i}{k_1} a_{\ell m}^{\mathrm{I}} \nabla \times j_\ell(k_1 r) \mathbf{X}_{\ell m}(\hat{\mathbf{r}}) \right\} , \tag{5.9a}$$

$$\mathbf{H}_{\mathrm{B}}^{\mathrm{I}}(\mathbf{r}) = \frac{1}{Z_1} \sum_{\ell m} \left\{ a_{\ell m}^{\mathrm{II}} j_\ell(k_1 r) \mathbf{X}_{\ell m}(\hat{\mathbf{r}}) - \frac{i}{k_1} b_{\ell m}^{\mathrm{I}} \nabla \times j_\ell(k_1 r) \mathbf{X}_{\ell m}(\hat{\mathbf{r}}) \right\} . \tag{5.9b}$$

The expansion coefficients $a_{\ell m}^{\mathrm{II/I}}$ and $b_{\ell m}^{\mathrm{II/I}}$ can be determined by matching the total fields inside and outside the NP, and are provided in Appendix F.2.2.

CL measurements are simulated with the CL probability, expressing the probability of collecting a photon of energy $\hbar\omega$ in the far field. It can be calculated by integrating the energy outflow of the induced fields over the surface of a sphere of radius $r \to \infty$ and solid angle Ω, as

Box 5.2 | CL probability

$$\Gamma_{\mathrm{CL}}(\omega) = \frac{r^2}{\pi\hbar\omega}\int d\Omega\,\mathrm{Re}\big\{\mathbf{E}^{\mathrm{II}}_{\mathrm{ind}}(\mathbf{r},\omega)\times\mathbf{H}^{\mathrm{II}*}_{\mathrm{ind}}(\mathbf{r},\omega)\big\}\cdot\hat{\mathbf{r}}\,. \tag{5.10}$$

By inserting Eqs. (5.8) into Eq. (5.10), we obtain

$$\Gamma_{\mathrm{CL}}(\omega) = \frac{1}{\pi\hbar\omega Z_0 k_0^2}\sum_{\ell m}\left\{\left|b^{\mathrm{II}}_{\ell m}\right|^2 + \left|a^{\mathrm{II}}_{\ell m}\right|^2\right\}. \tag{5.11}$$

Apart from the radiative energy dissipation, part of the energy transferred to the optical modes of the NP is absorbed in the material, due to its intrinsic loss mechanisms. The total energy lost is determined by the work done by the electron moving against the induced field along the entire electron trajectory. Then the (EEL) probability of the electron losing energy $\hbar\omega$ is given by

Box 5.3 | EEL probability

$$\Gamma_{\mathrm{EEL}}(\omega) = \frac{e}{\pi\hbar\omega}\int dt\,\mathrm{Re}\big\{\mathrm{e}^{-i\omega t}\,\mathbf{v}\cdot\mathbf{E}_{\mathrm{ind}}(\mathbf{r}_e,\omega)\big\}\,. \tag{5.12}$$

The integral in Eq. (5.12) can be decomposed into three terms (see details in Appendix F.2.2)

$$\Gamma_{\mathrm{EEL}}(\omega) = \Gamma_{\mathrm{bulk}}(\omega) + \Gamma_{\mathrm{surf}}(\omega) + \Gamma_{\mathrm{Begr}}(\omega)\,. \tag{5.13}$$

Here, Γ_{bulk} is associated with the bulk modes of the unbound medium, reduced by the Begrenzung term Γ_{Begr} that accommodates the influence of the boundary [36]. The Γ_{surf} term contains the contribution from modes excited by the part of the electron trajectory situated outside the NP. The terms entering Eq. (5.13) are obtained by the following expressions

$$\Gamma_{\mathrm{bulk}}(\omega) = \frac{e^2 z_0}{2\pi^2\varepsilon_0\hbar v^2}\mathrm{Im}\left\{\frac{1}{\gamma_0^2}\ln\left(\left[\frac{q_c\gamma_0 v}{\omega}\right]^2+1\right) - \frac{1}{\gamma_1^2\varepsilon_1}\ln\left(\left[\frac{q_c\gamma v}{\omega}\right]^2+1\right)\right\}, \tag{5.14a}$$

$$\begin{aligned}\Gamma_{\mathrm{surf}}(\omega) = \frac{e}{\pi\hbar\omega}\mathrm{Re}\sum_{\ell m}\Bigg\{&\frac{K_m\left(\omega b/[v\gamma_0]\right)}{ik_0\sqrt{\ell(\ell+1)}}\left[m b^{\mathrm{II}}_{\ell m}\mathscr{M}^*_{\ell m} - a^{\mathrm{II}}_{\ell m}\frac{\mathscr{N}^*_{\ell m}}{\beta\gamma_0}\right]\\ &-\int_{-z_0}^{z_0} dz\,\frac{\mathrm{e}^{-i\omega z/v}}{\sqrt{\ell(\ell+1)}}\left[m b^{\mathrm{II}}_{\ell m} h^+_\ell(k_0 r) Y^m_\ell(\theta,0) - \frac{a^{\mathrm{II}}_{\ell m}}{k_0 b}\left\{\mathscr{H}^+_{\ell m}(k_0 z) + \mathscr{H}^-_{\ell m}(k_0 z)\right\}\right]\Bigg\}, \end{aligned}\tag{5.14b}$$

and

$$\Gamma_{\mathrm{Begr}}(\omega) = \frac{e}{\pi\hbar\omega}\mathrm{Re}\sum_{\ell m}\int_{-z_0}^{z_0} dz \frac{\mathrm{e}^{-i\omega z/v}}{\sqrt{\ell(\ell+1)}}\Bigg[m b^{\mathrm{I}}_{\ell m} j_\ell(k_1 r) Y_\ell^m(\theta,0) - \frac{a^{\mathrm{I}}_{\ell m}}{k_1 b}\left\{\mathscr{I}^{+}_{\ell m}(k_1 z) + \mathscr{I}^{-}_{\ell m}(k_1 z)\right\}\Bigg]. \tag{5.14c}$$

It is important to note here that the expansion coefficients $a/b^{\mathrm{II/I}}_{\ell m}$ entering Eqs. (5.11) and (5.14) contain a correction that removes the contribution originating from the direct field of the electron. The exact expressions are given by Eqs. (F.99) in the Appendix.

Furthermore, we mention that the EELS decomposition of Eq. (5.13) introduces a free parameter: assuming that the electron deposits transverse (with respect to the electron trajectory) momentum q upon losing energy $\hbar\omega$ to excite an optical mode, q_{c} denotes the maximum transverse momentum collected. In an experiment, the momentum cutoff is established by the half-aperture collection angle ϕ_{out} of the microscope spectrometer, as

$$\hbar q_{\mathrm{c}} \approx \sqrt{(m_e v \phi_{\mathrm{out}})^2 + (\hbar\omega/v)^2}, \tag{5.15}$$

where m_e is the electron mass. We may freely select the q_{c} value, ensuring that it agrees with the typical collection angle values found in scanning TEM (STEM) setups, which are on the order of a few mrad. Naturally, this introduces a degree of arbitrariness in the EEL spectra, as different values of q_{c} result in different peak intensities at the BP energy.

5.2 CL and EELS of Metallic Nanoparticles

Using the formulas derived in the previous section, here we examine the response of a metallic NP excited by a swift electron. In what follows, we consider a nanosphere of radius $R = 75$ nm, characterized by the Drude dielectric function of Eq. (2.54), with plasma energy $\hbar\omega_{\mathrm{p}} = 5$ eV and damping rate equivalent to $\hbar\tau^{-1} = 50$ meV, and relative permeability $\mu = 1$.

Figure 5.2 shows the scattering spectrum (panel a) under plane-wave illumination, and the CL and EELS probability (panels b and c, respectively) computed for electron velocity $v = 0.33c$ (energy ≈ 30 keV) and impact parameter $b = 35$ nm. The scattering spectrum of Fig. 5.2a exhibits three peaks, associated with the radiative decay of LSP resonances. The features correspond to the contributions of the first three ($\ell = 1, 2, 3$) electric-type modes, while higher-order ($\ell > 4$) modes contribute negligible to the spectrum. As implied by Eq. (5.11), the total photon emission probability can also be analyzed into contributions of pairwise orthogonal electric and

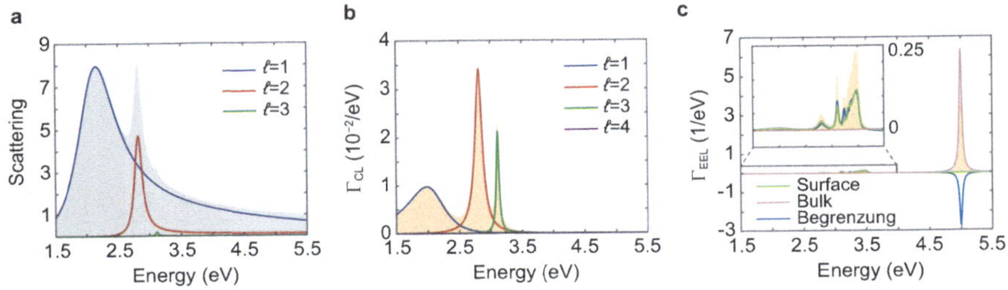

Fig. 5.2 Comparison of scattering, CL and EEL spectra of metallic NPs. a Normalized scattering cross section of a Drude NP of radius $R = 75$ nm under plane-wave excitation (gray shaded area). The spectrum is decomposed into the contributions of the ED (blue curve), EQ (red curve), and EO modes (green curve). **b** CL probability (yellow shaded area) decomposed into the contributions from the first four multipoles (blue, red, green, and violet curves). **c** EEL probability (yellow shaded area) analyzed into the surface, bulk, and Begrenzung terms (blue, green and pink curves), where the inset zooms in the 1–4 eV energy window. The bulk term is given by Eq. (5.14a) for $q_c = 0.71\,\mathrm{nm}^{-1}$. In panels **b** and **c** we consider electron velocity $v = 0.33c$ (energy ≈ 30 keV) and impact parameter $b = 35$ nm, and truncate the multipole expansion at $\ell_{\max} = 63$

magnetic multipoles of order ℓ. Figure 5.2b shows that the CL spectrum is composed of the contributions of the first four electric modes, arising at approximately 2, 2.8, 3.1, and 3.2 eV. Compared to the scattering cross section of panel a, the CL spectrum displays similar information. This is somewhat anticipated, since both calculations rely on the collection of far-field emission. Nevertheless, we do observe two notable distinguishing features in the CL spectrum of Fig. 5.2b. Firstly, the EQ and EO modes are excited more efficiently by the electron source, whereas in the scattering spectrum of panel a it is the ED mode that radiates more efficiently. The relative peak intensities in the CL and EEL spectra depend highly on the impact parameter, which dictates the arrangement of the polarization charges in the medium [38]. Secondly, we observe a slight redshift of the ED mode in CL due to retardation, owing to the fact that the velocity of the electron is only a fraction of the speed of light.

In Fig. 5.2c we calculate the EEL spectrum, decomposed as described in Eq. (5.13). In the zoom-in area of the figure we observe the excitation of numerous higher-order multipoles, manifesting as sharp peaks at energies reaching up to 3.5 eV. As the multipole order increases, the wavelength of the corresponding mode decreases, causing higher-order modes to experience the curved surface of the NP as progressively flatter. Consequently, they accumulate at the energy corresponding to that of SPPs at a planar interface, i.e. at $\hbar\omega_{\mathrm{spp}} = \hbar\omega_{\mathrm{p}}/\sqrt{2} \approx 3.5$ eV. Beyond the SPP energy, the spectrum features a prominent peak at the BP energy $\hbar\omega_{\mathrm{bp}} = \hbar\omega_{\mathrm{p}} = 5$ eV, stemming from the excitation of BPs in the volume of the NP. Since BPs are longitudinal modes, they cannot outcouple to far-field radiation and are thus detectable only with EELS. At the same energy, we observe the expected negative peak associated with the Begrenzung term, which reduces the BP peak in the total EEL probability [36].

It is worth acknowledging that the EEL spectrum of Fig. 5.2c is not fully converged. In principle, by modeling the electron beam as a point charge we allow for

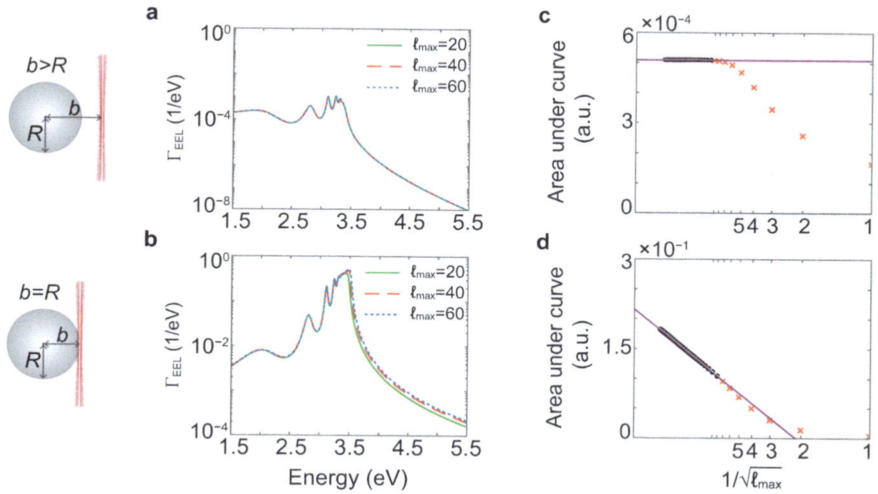

Fig. 5.3 Convergence of EELS for aloof and grazing electron trajectories. a, b EEL probability calculated for **a** an aloof electron trajectory with $b = 125$ nm, and **b** a grazing trajectory with $b = 75$ nm, and for selected values of $\ell_{\max}$, as denoted in the labels. **c, d** Area under the curve of the EEL spectra of panels **a** and **b**, respectively, versus $1/\sqrt{\ell_{\max}}$. In both panels, the purple line is fitted to the data points marked as black bullets, while the red marks correspond to data excluded from the fitting. The vertical gray lines serve as guides to the eye for the positions of $\ell_{\max} = 1, 2 \ldots 9$

an infinite number of high-order multipoles to be excited, so the accuracy of the value at $\hbar\omega_{\text{spp}}$ boils down to the number of multipoles we are able to include in the summation of Eq. (5.13) before we exhaust our computational resources, compared to how quickly the summation converges. The convergence study presented in Figs. 5.3 and 5.4 reveals that this depends on the impact parameter. Figure 5.3a and b show the EEL spectrum of the same system as before, now for an aloof (panel a) and a grazing electron trajectory (panel b), for varying values of the multipole cutoff $\ell_{\max}$, indicated by curves of different colors. The EEL spectrum of the aloof trajectory does not change, as we factor in additional multipoles, meaning that it is already converged at $\ell_{\max} = 20$. In contrast, in the grazing case, even up to $\ell_{\max} = 60$, the spectrum changes as more multipoles are taken into account, specifically around $\hbar\omega_{\text{spp}}$ where those modes pile up; below this energy no contribution is added, since different multipoles are orthogonal to each other. This suggests that tracking the integrated total EEL spectrum (area under the curve) can serve as a proxy for convergence.

In Fig. 5.3c and d we plot the area under the EEL curve versus $1/\sqrt{\ell_{\max}}$. Indeed, in panel c we observe that the curve converges rapidly, already for $\ell_{\max} = 10$, while the addition of more multipoles does not change the final value, indicated by the flattening of the fitting curve (purple line). The result is in agreement with the exponential convergence of the asymptotics of the Hankel functions of Eqs. (5.8). In contrast, Fig. 5.3d shows a slow convergence in the case of the grazing trajectory,

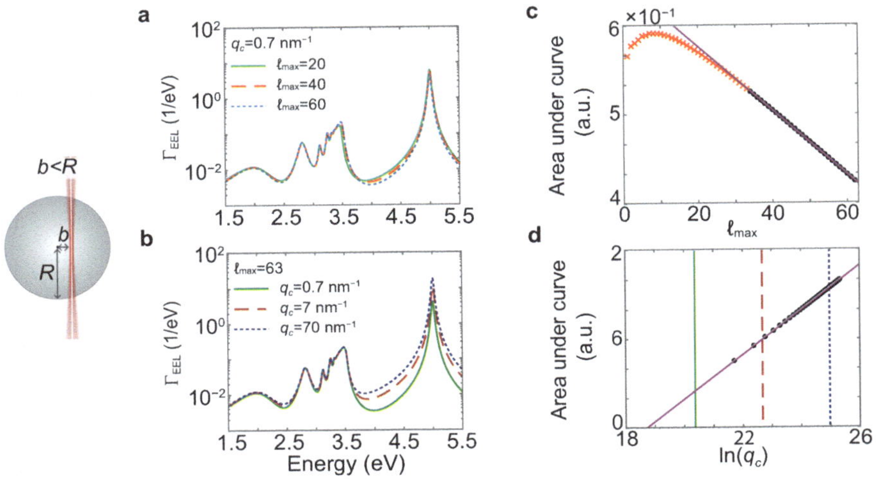

Fig. 5.4 Convergence of EELS for penetrating electron trajectories. a, b EEL probability calculated for a penetrating electron trajectory with $b = 10\,\text{nm}$, **a** for varying $\ell_{\max}$ and fixed $q_c = 0.7\ \text{nm}^{-1}$, and **b** for varying q_c and fixed $\ell_{\max} = 63$, as denoted in the labels. **c** Area under the curve of the EEL spectra of panel **a** versus $\ell_{\max}$, and **d** Area under the curve of the EEL spectra of panel **b** versus $\ln(q_c)$, with the vertical lines showing the selected q_c values shown in panel **b**, according to the color coding. In panels **c** and **d**, the purple line is fitted to the data points marked as black bullets, while the red marks correspond to data excluded from the fitting

which breaks down to a square-root law. By extrapolation of the fitting curve (purple line), it is evident that the EEL spectrum is approximately 85% converged, while the area missing from the converged value corresponds to the higher-order modes accumulating at the SPP energy, according to Fig. 5.3b. Finally, we note that the curve in panel c also features a square-root convergence before it flattens off. The precise value of $\ell_{\max}$ where this transition occurs increases as the impact parameter b tends to the size of the radius R, namely as higher order modes are excited.

For the penetrating trajectory shown in Fig. 5.4 we observe again a slow convergence around the SPP accumulation point. This convergence behavior is not obvious in our convergence plots due to the divergence of the BP peak. The source of this divergence can be traced back to the decomposition of the EEL probability into two competing terms, the bulk and the Begrenzung. As illustrated in Fig. 5.4c and d, the two terms lead to divergences of opposite sign; the negative Begrenzung term follows a linear divergence with increasing multipole order ($\ell_{\max}$), while the positive bulk term diverges logarithmically with increasing momentum cutoff (q_c) [see Eqs. (5.14a) and (5.14c)]. Since computational limitations can never allow us to extend both $\ell_{\max}$ and q_c to infinity, once one of two parameters is truncated to a specific cutoff value, the other must be adjusted accordingly. Figure 5.4a and b illustrate how the different values of $\ell_{\max}$ and q_c, respectively, impact the spectra spanning from the SPP to the BP energy.

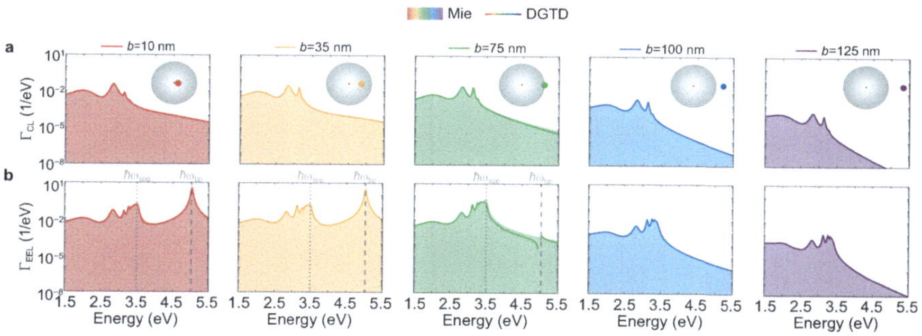

Fig. 5.5 Analytic and DGTD calculation of the CL and EEL probability for $v = 0.33c$. **a** CL and **b** EEL probability in the interaction between the metallic NP and an electron traveling with velocity $v = 0.33c$ (energy $\approx 30\,\mathrm{keV}$), considering 5 different impact parameters, as denoted in the labels and illustrated in the insets of each panel. The shaded areas correspond to results of the analytic Mie-type calculation, whereas the solid lines show the corresponding DGTD simulations. In the three leftmost panels of **b** the vertical gray dotted and dashed lines trace the energy of the SPP ($\hbar\omega_{\mathrm{spp}} \approx 3.5\,\mathrm{eV}$) and the BP mode ($\hbar\omega_{\mathrm{bp}} = 5\,\mathrm{eV}$), respectively. We use $\ell_{\max} = 63$ and $q_c = 0.7\ \mathrm{nm}^{-1}$ in the calculations

In contrast to our modeling of the electron as a point charge, actual electron beams in electron spectroscopies have a finite width, which in fact determines the smallest wavelength resolvable. Therefore, at the SPP energy, rather than a sharp high-intensity feature, a much broader resonance is typically observed. The suppression of the peak and, essentially, the smearing of higher-order modes can be replicated in the analytic theory, once we take the nonlocal response of the material into account; this is a particularly easy task in Mie theory [39–41]. Nonlocal effects manifest in the CL and EEL spectra as linewidth broadenings, due to increased damping, and energy shifts of the plasmonic modes, by an amount that depends on the mode order [42]. As a result, the amplitude of each mode individually is suppressed, and their overlap at $\hbar\omega_{\mathrm{spp}}$ is reduced.

5.2.1 Comparison with DGTD

Having a clear picture of the origin of all spectral features of the metallic NP discussed in Fig. 5.2b and c, in Figs. 5.5 and 5.6 we compare the CL and EEL probability as calculated employing the analytic (Mie) theory and with the DGTD method[1], as described in Sect. 2.4.2.[2] A notable difference between the analytic theory and the DGTD simulations lies in the modeling of the excitation source. In DGTD the electron is modelled with a Gaussian charge distribution

[1] All DGTD simulations were performed by Wenhua Zhao, and they are presented in this thesis with her permission.

[2] For technical details about the implementation of electron beams in DGTD we refer the reader to Preprint A and the references therein.

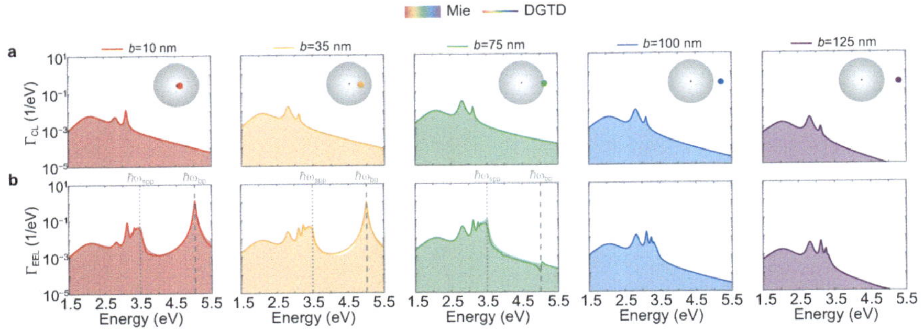

Fig. 5.6 Analytic and DGTD calculation of the CL and EEL probability for $v = 0.7c$. **a** CL and **b** EEL probability in the interaction between the metallic NP and an electron traveling with velocity $v = 0.7c$ (energy $\approx 200\,\text{keV}$), considering 5 different impact parameters, as denoted in the labels and illustrated in the insets of each panel. The shaded areas correspond to results of the analytic Mie-type calculation, whereas the solid lines show the corresponding DGTD simulations. In the three leftmost panels of **b** the vertical gray dotted and dashed lines trace the energy of the SPP ($\hbar\omega_{\text{spp}} \approx 3.5\,\text{eV}$) and the BP mode ($\hbar\omega_{\text{bp}} = 5\,\text{eV}$), respectively. We use $\ell_{\max} = 63$ and $q_c = 19\ \text{nm}^{-1}$ in the calculations

$$\rho(\mathbf{r}) = -\frac{e}{\sigma_e^3\sqrt{\pi^3}}\text{e}^{-r^2/\sigma_e^2}\,, \tag{5.16}$$

of width $\sigma_e = 5\,\text{nm}$, which is compatible with spot sizes used in CL experiments (see e.g. Publication D). This value is expected to produce non-negligible differences for grazing and penetrating electron beams, as we discuss below.

In Fig. 5.5 we examine the agreement between the two calculations testing five impact parameters, that range from $b = 125\,\text{nm}$ to $10\,\text{nm}$, corresponding to aloof (violet and blue spectra), grazing (green), and penetrating (yellow and red) electron trajectories. All in all, the DGTD approach reproduces the positions of the LSP modes and the corresponding CL probabilities with remarkable accuracy. As depicted in Fig. 5.5a, for aloof and penetrating electron trajectories we observe an excellent agreement in the CL spectra, with a relative error approximately of 1%, as shown in Table 5.1. In passing, here we define the relative error as

$$\text{Rel. error} = \frac{\int d\omega\,|\Gamma_{\text{Mie}}(\omega) - \Gamma_{\text{DGTD}}(\omega)|}{\int d\omega\,|\Gamma_{\text{Mie}}(\omega)|} \cdot 100\%\,, \tag{5.17}$$

where Γ indicates the curves under comparison, and the integrals span from 1.5 to 5.5 eV. The largest error is obtained when the electron passes at the rim of the NP, at $b = R$ (middle panel in Fig. 5.5a). This case highlights a crucial limit in the capabilities of the DGTD method; owing to the finite width of the electron beam, half of the Gaussian charge density distribution lies inside the NP, while the other half extends outside, resulting in numerical inconsistencies.

In the EEL spectra of Fig. 5.5b, we find an excellent agreement for aloof electron trajectories. As expected, we find the highest error for grazing trajectories (middle

Table 5.1 Relative error between the analytic and DGTD calculation of the CL and EEL probability for varying impact parameter

$v = 0.33c$			$v = 0.7c$		
b (nm)	Γ_{CL} (%)	Γ_{EEL} (%)	b (nm)	Γ_{CL} (%)	Γ_{EEL} (%)
10	1.26	9.75	10	0.59	14.67
35	1.06	6.17	35	0.55	11.10
75	8.23	28.87	75	3.52	19.75
100	1.06	1.02	100	1.14	0.73
125	1.07	1.93	125	1.15	1.14

panel in Fig. 5.5b), for the reasons mentioned above. In this case, the numerical EEL curve computed using DGTD features a numerical artifact at the BP energy, marked by the gray dashed line, as a consequence of the partial excitation of the bulk mode. There can be seen significant disparities between the two methods in the EEL spectra for both grazing and penetrating electron trajectories (three leftmost panels in Fig. 5.5b) between the SPP and the BP energy, indicated by the gray dotted and dashed lines, accordingly. This discrepancy translates into the high relative errors presented in Table 5.1. The reason is that the finite width of the electron beam introduces an additional spatial scale. For very large distances between the electron and the NP (i.e., $b - R \gg \sigma_e$), the EM fields created by the source resemble those of a point charge; therefore, we obtain an excellent agreement with analytic results. This is not the case, once this distance becomes comparable to the beam width. At the same time, the value of the width is commensurate with other relevant length scales of the computational domain, e.g. the mesh element size,[3] and distinct choices for σ_e lead to substantial differences in the spectra.

Lastly, there is an additional source of error to consider, which arises at the BP peak, due to the rather arbitrary value assigned to the transverse momentum cutoff q_c in the analytic approach. In contrast, in the DGTD implementation there exists an internal limit, linked to σ_e. Consequently, we consistently find greater discrepancies in EELS compared to CL, and greater relative errors for grazing and penetrating electron trajectories compared to aloof.

For completeness, in Fig. 5.6, we compute the CL and EEL probability considering the same system as in Fig. 5.5, but for a higher electron velocity $v = 0.7c$ (energy 200 keV), a value that is more appropriate for EELS. Figure 5.6 displays no qualitative differences when compared to Fig. 5.5, apart from a larger deviation in EELS at the valley between the SPP (gray dotted line) and the BP energy (gray dashed line) for the penetrating trajectories, also reflected in the high corresponding errors presented in Table 5.1.

[3] The surface and interior of the NP are more finely discretized than the surrounding medium.

5.3 CL and EELS of Dielectric Nanoparticles

The interaction of high-index dielectrics with penetrating electron beams is particularly interesting. It should be clear by now that electrons can probe efficiently higher order modes, while also offering the ability to selectively excite resonances by adjusting the impact parameter [9]. These properties enable us not only to study individual modes, but also to actively tune effects associated with their overlap, for instance for the generation of circularly polarized light [43] or the mapping of anapoles [44]. Furthermore, since they are characterised by a high positive permittivity, they enable the emission of CR when the electron propagates faster than the phase velocity of light in the medium, i.e. when the condition $\mathrm{Re}\{\varepsilon\}v^2 > c^2$ is satisfied.

To study the excitation of Mie resonances with electrons, we consider a Si NP of radius $R = 75$ nm, and permittivity given by the experimental values provided in Ref. [45]. The dielectric function of Si, and the extinction cross section under plane-wave illumination of Si nanospheres of varying radii, are presented in Fig. 2.6 of Chap. 2. Revisiting the middle panel of Fig. 2.6c, we indeed observe the large number of electric and magnetic modes sustained, as well as significant overlap between them at certain energies. Figure 5.7 presents the CL and EEL probability in the interaction

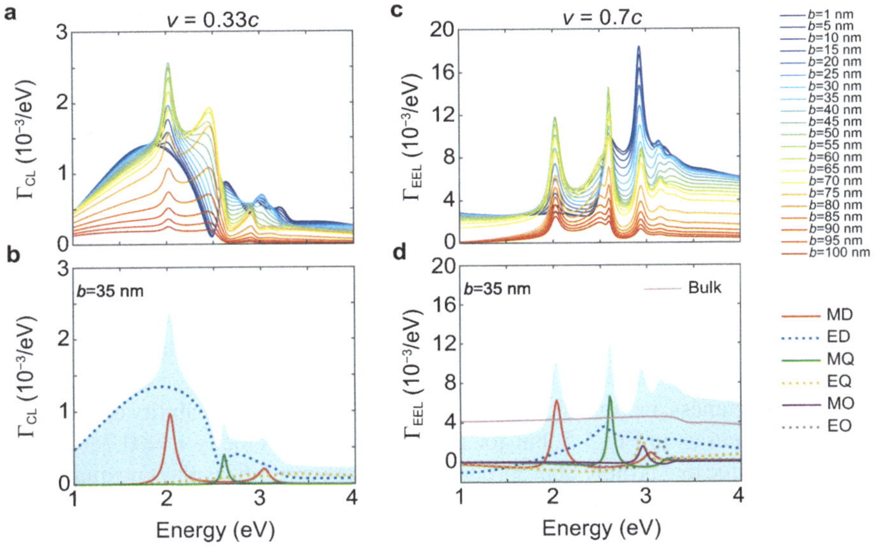

Fig. 5.7 CL and EEL spectra of a Si NP. **a** CL probability of a Si NP of radius $R = 75$ nm excited by an electron traveling with $v = 0.33c$ (energy $\approx$ 30 keV) for various impact parameters. **b** Total CL probability (cyan shaded area) at $b = 35$ nm, and contributions from the MD, ED, MQ, EQ, MO, and EO modes (solid and dashed curves). Higher-order modes contribute negligibly to the CL spectrum. **c** EEL probability of the same Si NP excited by an electron traveling with $v = 0.7c$ (energy $\approx$ 200 keV) for various impact parameters. **d** Total EEL probability (cyan shaded area) at $b = 35$ nm, and contributions from the MD, ED, MQ, EQ, MO, and EO modes (solid and dashed curves). The bulk contribution Γ_{bulk} of the unbound medium is plotted with the pink line

between the Si NP and an electron of velocity $v = 0.33c$ (energy $\approx$30 keV) and $v = 0.7c$ (energy $\approx$ 200 keV), respectively. Panels a and c showcase the sensitivity of the spectra on the impact parameter: as we scan various values ranging from $b = 1$ nm to 100 nm, encompassing both penetrating and aloof trajectories, we observe substantial qualitative differences. When considering aloof electron trajectories, the CL spectra of panel a resemble the scattering cross-section spectrum depicted in Fig. 2.6c (middle panel). However, as the electron beam approaches the center of the NP, the peaks associated with the Mie resonances gradually diminish until, for $b < 10$ nm, we only observe an oscillatory trend. Clearly, the peaks and dips of said oscillation are unrelated to the Mie resonances, since the latter are eigenmodes of the structure, and thus their eigenfrequencies are independent of the excitation. Only small shifts can emerge, due to background contributions from evanescent tails of overlapping modes or other loss mechanisms that might occur. The multipole decomposition shown in panel b for $b = 35$ nm, suggests that this oscillatory behavior originates from the ED mode. Apart from this, three sharp peaks can be discerned, corresponding to the first and second radial order of the MD mode (red curve) and the MQ mode (green curve), appearing at the expected energies, according to the middle panel of Fig. 2.6c. As we discuss thoroughly in the next section, the observed oscillation stems from the interference of TR emanating each time the electron crosses the air–NP interface. The effect becomes particularly evident for small impact parameters, owing to the low excitation efficiency of the magnetic multipoles. In fact, when the electron beam traverses the center of the structure ($b = 0$) the induced current loops cancel out due to the symmetry of the structure, resulting in a zero contribution in the CL probability.

In the EEL spectra at $v = 0.7c$ in Fig. 5.7c we once again observe substantial differences for varying impact parameter. However, the previously observed oscillatory behavior of the ED mode is absent in this case, indicating that TR emission is not the primary loss mechanism at high electron velocities. Here we observe three pronounced peaks at energies 2.02, 2.59 and 2.94 eV which correspond to the MD, MQ and MO mode respectively, according to panel d. Between approximately 50 nm $< b <$ 100 nm we observe a broad shoulder at 2.5 eV, that corresponds to the ED mode. For energies higher than 3 eV, the EEL spectra of c and d show a considerable contribution from higher order electric and magnetic multipoles, but they do not appear as pronounced peaks due to their overlap and the overall high contribution from bulk losses (pink line in panel d).

Notably, for small impact parameters the entire EEL spectrum exhibits a blueshift, which can be discerned more clearly in Fig. 5.8a, where we plot the contribution of the ED mode exclusively, considering the same electron trajectories as earlier. By tracing the first peak, emerging at 2.5 eV, we observe a blueshift of the order of 100 meV. Nevertheless, no blueshift emerges in the EEL spectra obtained for a lower electron velocity $0.2c$ (energy $\approx$ 11 keV) in Fig. 5.8b. At this velocity, the ED mode couples efficiently to the electron field only for small impact parameters. Beyond 2.6 eV the energy lost from the electron is mainly transferred to interband transitions (ITs) (see the onset of high values in $\mathrm{Im}\{\varepsilon\}$ in Fig. 2.6b). The fact that the blueshift is featured only at high electron velocities is a strong indication that it is a consequence of CR; this is further corroborated by the color maps presented in Fig. 5.8c–e. Panels

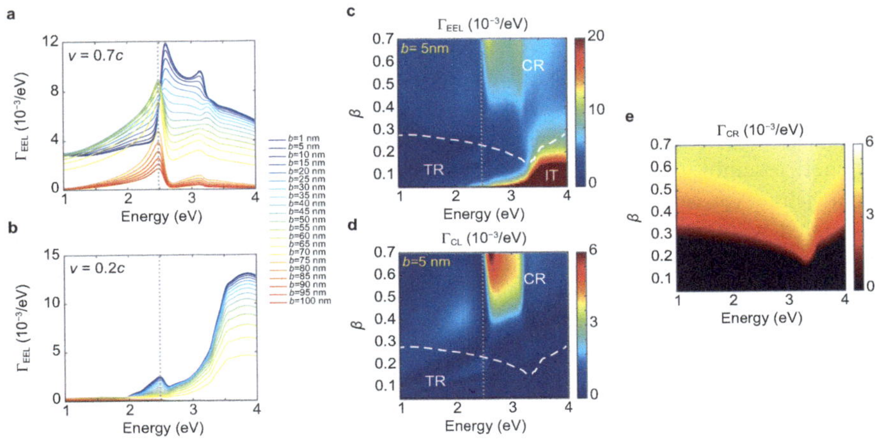

Fig. 5.8 **Interpretation of EEL spectra of Si NPs. a, b** Contribution of the ED mode to the EEL probability of a Si NP of radius $R = 75$ nm excited by an electron traveling with **a** $v = 0.7c$ (energy ≈ 200 keV) and **b** $v = 0.2c$ (energy ≈ 11 keV), for various impact parameters. **c** Contribution of the ED mode to EELS for $b = 5$ nm versus the kinematic factor $\beta = v/c$. **d** Contribution of the ED mode to CL for $b = 5$ nm versus the kinematic factor $\beta = v/c$. In panels **c** and **d**, the white dashed lines indicate the limit $\text{Re}\{\varepsilon\}\, v^2 > c^2$, above which CR can be generated. The labels TR, CR and IT denote the energy-velocity windows where the corresponding mechanism is the dominant energy loss channel. In panels **a**–**d** the vertical dotted line indicates the energy of the ED mode, as predicted by Mie theory. **e** CR probability versus the kinematic factor $\beta = v/c$ for an electron traveling length $d = 2z_0$ in bulk Si

c and d show the contribution of the ED mode to the EEL and CL probability, respectively, for $b = 5$ nm, and velocities ranging from $0.05c$ to $0.7c$. The white dashed line marks out the region above which CR can be effectively generated, i.e. where $\text{Re}\{\varepsilon\}\, \beta^2 > 1$. In panel e we calculate the probability of Cherenkov emission from the bulk of the dielectric medium for an electron traveling distance $d = 2z_0$ according to the Frank-Tamm formula [46]

$$\Gamma_{\text{CR}}(\omega, \beta) = \frac{de^2}{4\pi\varepsilon_0\hbar c^2}\text{Re}\left\{1 - \frac{1}{\beta^2\varepsilon(\omega)}\right\}. \tag{5.18}$$

Overall, CR appears as a background contribution for velocities above $0.3c$, admitting its highest value at around 3.32 eV. Ultimately, in Fig. 5.8c we can distinguish three regimes: firstly, at low energies and velocities the spectra are dominated by interference TR signals (labeled as TR). Secondly, at low velocities, and above 2.6 eV, energy is lost to the ITs in Si (labeled as IT). Finally, for $v > 0.3c$ CR can be emitted. The only region where Mie resonances can be clearly discerned, albeit blueshifted, is the latter. In the CL spectra of Fig. 5.8d, we can identify the regions of only radiative loss, i.e. TR and CR. The transition from the TR to the CR regime emitted from a dielectric NP excited by an electron traversing its center, has already been reported by Pogorzelski and Yeh [33], who demonstrated that the far-field spectra at low velocities, and below the Cherenkov limit, are characterized by a dependence (for $b = 0$)

$$\Gamma_{\mathrm{CL}} \propto \frac{\sin(x)}{x}\,, \quad x = \frac{\omega R}{v}\,. \tag{5.19}$$

The following section takes a deeper look into the imprint of TR in the spectra of dielectric nanospheres.

5.3.1 *The Role of Transition Radiation*

It should be clear by now, that electron beams traversing small NPs can potentially provide more information compared to aloof trajectories. Apart from the optical modes of the structure, the interaction of fast electrons with the material can trigger a number of unrelated mechanisms, that act as sources of coherent radiation propagating to the far-field, and are eventually collected in CL measurements. Here we address one such mechanism: TR. We briefly mentioned in the previous section that TR is created when a fast electron crosses the interface between two media [47, 48]. Consider a fast electron impinging on a spherical NP, as depicted in Fig. 5.9. As the electron advances towards the NP surface, so does the image charge created inside the NP, until, at contact, the two charges collapse, resulting in the creation of an effective radiating electric dipole; TR refers exactly to this dipole-like emission. In finite structures, like the spherical NPs examined here, a similar process takes place when the electron exits the NP, after a time delay $t_{\mathrm{d}} = d/v$. The TR signals can

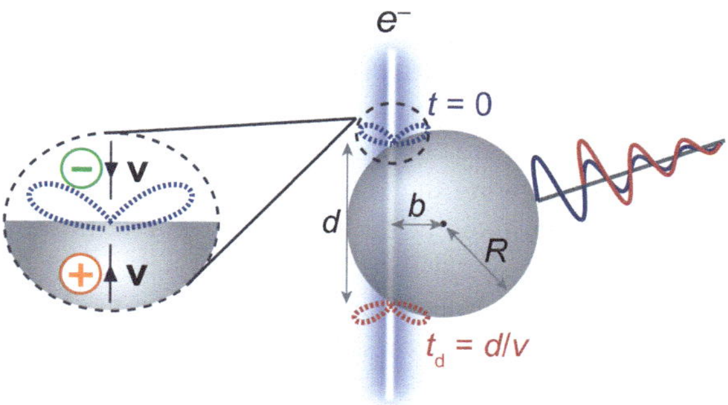

Fig. 5.9 **TR generation and interference**. Schematic illustration of a fast electron traversing a spherical NP of radius R, traveling with velocity v a distance $d = 2z_0$ inside it. When the electron crosses the NP–air interface it cancels its mirror image, creating an effective radiating dipole. TR is generated when the electron enters the NP, at time $t = 0$, and when it exits, at $t_{\mathrm{d}} = d/v$. Depending on the time delay t_{d} between the two events, the generated TR signals can interfere constructively or destructively

interfere constructively or destructively, depending on the time delay between these two events, which in turn is dictated by the electron velocity v and the length d of its path inside the NP. In this section we present a comparison between theoretical CL spectra of Si NPs, calculated using the analytic theory presented in Sect. 5.1, with the experimental measurements published in Publication D.[4] Here we focus on the theoretical aspects of this work, specifically the role of TR in the spectra and its interplay with the Mie resonances sustained by Si nanospheres.[5]

The CL measurements were performed in the setup shown schematically in the left-hand sketch of Fig. 5.10c: a Si nanosphere, situated on a thin Si_3N_4 membrane, is excited by an electron beam of energy 30 keV, and the emitted light is then collimated and gathered using a parabolic mirror. Figure 5.10a and b show the CL measurement and the simulated CL probability, respectively, versus photon energy and for varying impact parameter b, corresponding to either aloof or penetrating electron beams, while the horizontal dashed line denotes the grazing trajectory $b = R$. The color map of panel a concerns a NP of radius $R = 62$ nm (see SEM image on the right of Fig. 5.10a), for which the modal characterization should be straightforward. With vertical dashed lines we indicate the energies where the ED and MD modes are expected to appear from the extinction cross section obtained with Mie theory. Two resonances manifest clearly in the spectra for aloof electron trajectories; a pronounced MD mode at about $\hbar\omega = 2.35$ eV, and a broader and less intense ED mode around 2.75 eV. However, for penetrating electron beams, and as the impact parameter is reduced, while the MD mode remains at the same position until it vanishes at $b = 0$ (due to the symmetry reasons mentioned earlier), the feature seemingly associated with the ED mode gradually shifts to lower energies. This is a manifestation of TR, whose condition of interference changes with the impact parameter that determines the time delay $t_{\rm d}$. Therefore, when penetrating beams are considered and TR-related signals become important, the Mie resonances are masked by peaks and dips associated with interference maxima and minima, respectively. This trend is in good agreement with the simulated CL emission in the upper half-space in panel b. Differences between theoretical extinction and experimental CL spectra can be ascribed to NP shape imperfections and the influence of the substrate [49, 50].

To ensure that these features are consistent with the interference of TR, we next modify the time delay by changing the velocity of the electrons while keeping the position of the electron beam fixed at the center of the NP. In the four panels of Fig. 5.10d, we collect the CL signal for electron energies 5, 10, 20, and 30 keV (solid curves). We observe broad peaks at energies where no Mie resonances are expected, as denoted by the green arrows, while, at the same time, dips appear at the energy where the ED mode should emerge, for electron energies 20 and 30 keV. These features are evident also in the simulated CL probability depicted with the

[4] The samples were fabricated by Hiroshi Sugimoto and Minoru Fujii, and the CL measurements were performed by Saskia Fiedler and Artyom Assadillayev. All experimental data are presented in this thesis with their permission.

[5] We refer the reader to Publication D for a more detailed account of the experimental setup and methods.

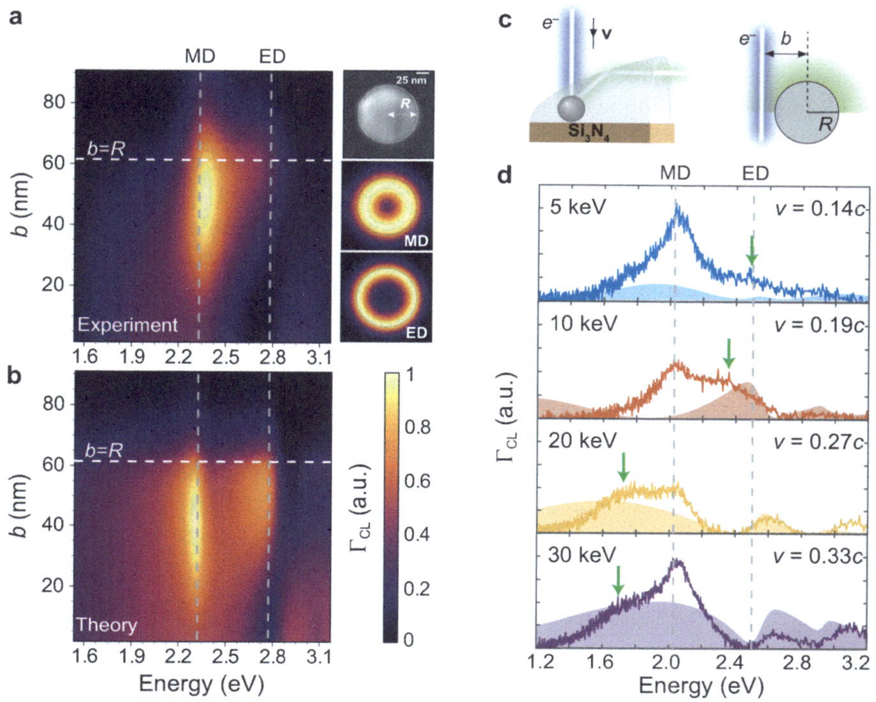

Fig. 5.10 Experimental and theoretical CL spectra of Si NPs. a Left: Experimental CL intensity of a Si nanosphere of radius $R = 62$ nm, as a function of energy and impact parameter b. The kinetic energy of the electron is 30 kV ($v \approx 0.33c$). The vertical dashed lines denote the energies of the MD and ED resonances predicted by analytic extinction calculations with Mie theory, and the horizontal indicates the grazing electron trajectory. Right: SEM image of a typical Si NP used in the experiments, and experimental CL maps at the MD and ED energies. **b** CL probability simulating the color map of panel **a**. Panels **a** and **b** share the same color scale and they are both normalized to the highest value obtained in each case. **c** Left: Schematic illustration of the experimental setup showing a Si NP, situated on a thin Si_3N_4 membrane, and excited by an electron beam. The emitted radiation is collimated and collected by a parabolic mirror. Right: Sketch of the theoretical description showing the electron passing at an impact parameter b, and the photon emission collected at the upper half-space (green-shaded area). **d** Experimental (solid curves) and simulated (shaded areas) CL spectra for an electron beam passing through the center of a Si nanosphere of radius $R = 75$ nm, for electron-beam energies ranging from 5 keV (upper panel) to 30 keV (lower panel). Vertical dashed lines denote the energies of the MD and ED modes as predicted by Mie theory. The green arrows indicate some of the spectral features that are not related to the Mie resonances of the NP

shaded areas in the same panels. Here, there is a large deviation between theory and experiment regarding the MD mode. As mentioned in the preceding section, the magnetic-type modes cannot be excited by beams passing close to the center of the NP for symmetry reasons. While this is reflected in the theoretical spectra, it is not the case in the experimental, because the electron beam does not follow strictly a straight-line trajectory, and eventually current loops can be formed in the NP.

Moreover, it is worth mentioning that for decreasing electron velocity, the non-recoil approximation invoked in the derivation of the CL probability becomes less valid. Regardless of the disagreement, the theoretical spectrum reproduces the positions of the TR interference minima and maxima observed in the experiment, and highlights that the oscillation resulting from this interaction dominates the spectra.

A clearer picture of the TR-related oscillation can be obtained by Fig. 5.11b, that summarizes in a single panel the theoretical spectra presented in Fig. 5.10d, for a larger energy window. The two effective dipoles created as a result of the collapse of the electron and its mirror image (see schematics in panel a) interfere, leading to CL signals that emerge as a superposition of NP resonances and interfering TR. For the four different time delays displayed, the CL probability in panel b is dominated by the resulting oscillation, that can be seen to redshift in energy for decreasing electron velocity. Trace, for instance, the first maximum that lies at $\hbar\omega = 1.95$ eV in 30 keV, to $\hbar\omega = 1.63$ eV at 20 keV and $\hbar\omega = 1.11$ eV at 10 keV. This intricate interaction is more evident in the CL color map of Fig. 5.11c, where we plot the CL spectra versus $\beta = v/c$. Here, the TR interference minima appear as straight-line valleys. In fact, the lines follow the relation predicted by Pogorzelski and Yeh [33]: the white dotted lines in panel c follow Eq. (5.19), $\cos(x)/x$ with $x = \omega R/v$ (here expressed with

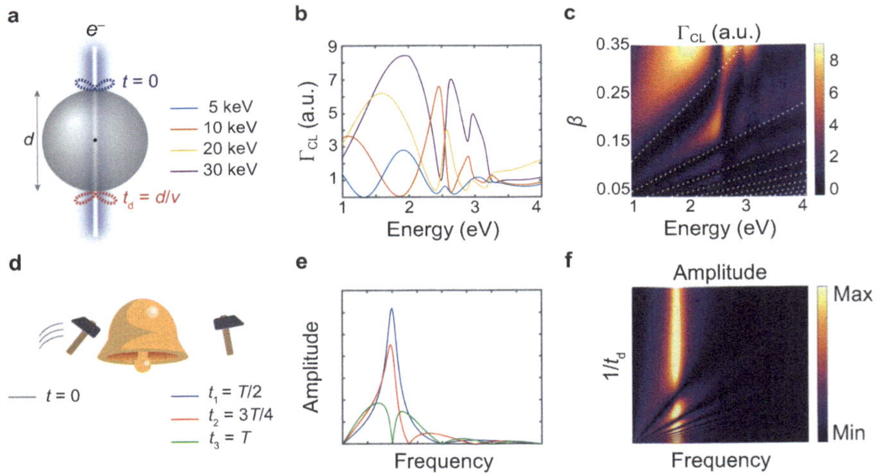

Fig. 5.11 **Acoustic analogue for interpreting the CL spectra. a** Emergence of the two effective radiating dipoles at the upper and lower surfaces of a NP traversed by an electron beam through its center. **b** CL probability simulations for a Si sphere with $R = 75$ nm, same as in Fig. 5.10d, following the same color coding for the different electron velocities. **c** Color map of the CL probability, as a function of the kinematic factor $\beta = v/c$. The dashed white lines trace the relation $\cos(x)/x$, where $x = \omega R/v$. **d** A bell hit by two hammers from two opposite sides, with a selection of three time delays t_i comparable to the period T of the generated acoustic waves. **e** Total amplitude of the acoustic wave generated at $t = 0$ and interfering with the delayed acoustic waves (blue, red and green lines) produced by the two hammers in the far field, in the frequency domain. **f** Color map of the amplitude as a function of frequency and the inverse of the time delay t_d

cosine in order to trace the minima, rather than the maxima) and accurately trace the valleys.

To obtain an intuitive understanding of the effects discussed above, let us first focus on the classical analogue depicted in Fig. 5.11d–f. Figure 5.11d shows a bell which, at time $t = 0$, is hit by a hammer on its left side, thereby generating an acoustic wave oscillating with period T. A second hammer strikes the bell on the opposite side, with some time delay t_d, launching a second wave that interferes with the initial one. In Fig. 5.11e we plot the amplitude of the total wave that reaches an observer (in this case listener) in the frequency domain, considering three different time delays for the second signal, corresponding to $T/2$ (blue line), $3T/4$ (red line), and T (green line). The frequency-domain signal displays an oscillatory behavior, with zero amplitude emerging whenever the conditions for destructive interference are satisfied. The amplitude color map of Fig. 5.11f shows, for each time delay, the locations of destructive interference minima at different frequencies, superimposed over the eigenfrequency of the oscillator, appearing as straight-line valleys with respect to the inverse time delay and the frequency.

In summary, when penetrating electron beams are considered, apart from the eigenmodes of the structure, other sources of coherent radiation may dominate the CL and EEL spectra, which is particularly evident in the optical response of dielectric NPs. Our findings suggest that the CL spectra of Si nanospheres are extremely sensitive to TR, thus care must be taken when assigning modal character to the spectral features observed. To avoid these effects, one can always choose to employ aloof electron beams only, as highly energetic as possible.

References

1. Bosman M, Keast VJ, Watanabe M, Maaroof AI, Cortie MB (2007) Nanotechnology 18:165505. https://doi.org/10.1088/0957-4484/18/16/165505
2. Nelayah J, Kociak M, Stéphan O, García de Abajo FJ, Tencé M, Henrard L, Taverna D, Pastoriza-Santos I, Liz-Marzán LM, Colliex C (2007) Nat Phys 3:348. https://doi.org/10.1038/nphys575
3. García de Abajo FJ (2010) Rev Mod Phys 82:209. https://doi.org/10.1103/RevModPhys.82.209
4. Ritchie RH (1957) Phys Rev 106:874. https://doi.org/10.1103/PhysRev.106.874
5. Watanabe H (1956) J Phys Soc Jpn 11:112. https://doi.org/10.1143/JPSJ.11.112
6. Yamamoto N, Araya K, García de Abajo FJ (2001) Phys Rev B 64:205419. https://doi.org/10.1103/PhysRevB.64.205419
7. Vesseur EJR, de Waele R, Kuttge M, Polman A (2007) Nano Lett 7:2843. https://doi.org/10.1021/nl071480w
8. Coenen T, Van De Groep J, Polman A (2013) ACS Nano 7:1689. https://doi.org/10.1021/nn3056862
9. Matsukata T, Matthaiakakis N, Yano T, Hada M, Tanaka T, Yamamoto N, Sannomiya T (2019) ACS Photon 6:2320. https://doi.org/10.1021/acsphotonics.9b00833
10. Lagos MJ, Trügler A, Hohenester U, Batson PE (2017) Nature 543:529. https://doi.org/10.1038/nature21699
11. Hage FS, Kepaptsoglou DM, Ramasse QM, Allen LJ (2019) Phys Rev Lett 122:016103. https://doi.org/10.1103/PhysRevLett.122.016103

12. Maciel-Escudero C, Konečná A, Hillenbrand R, Aizpurua J (2020) Phys Rev B 102:115431. https://doi.org/10.1103/PhysRevB.102.115431
13. Polman A, Kociak M, García de Abajo FJ (2019) Nat Mater 18:1158. https://doi.org/10.1038/s41563-019-0409-1
14. García de Abajo FJ, Di Giulio V (2021) ACS Photon 8:945. https://doi.org/10.1021/acsphotonics.0c01950
15. Egerton RF (2011) Electron energy-loss spectroscopy in the electron microscope. Springer Science & Business Media
16. Batson PE, Silcox J (1983) Phys Rev B 27:5224. https://doi.org/10.1103/PhysRevB.27.5224
17. Schmidt F-P, Ditlbacher H, Hohenester U, Hohenau A, Hofer F, Krenn JR (2012) Nano Lett 12:5780. https://doi.org/10.1021/nl3030938
18. Zhou X, Hörl A, Trügler A, Hohenester U, Norris TB, Herzing AA (2014) J Appl Phys 116:223101. https://doi.org/10.1063/1.4903535
19. Koh AL, Bao K, Khan I, Smith WE, Kothleitner G, Nordlander P, Maier SA, McComb DW (2009) ACS Nano 3:3015. https://doi.org/10.1021/nn900922z
20. Coenen T, Polman A (2012) Opt Express 20:18679. https://doi.org/10.1364/OE.20.018679
21. Barwick B, Park HS, Kwon O-H, Baskin JS, Zewail AH (2008) Science 322:1227. https://doi.org/10.1126/science.1164000
22. Barwick B, Flannigan DJ, Zewail AH (2009) Nature 462:902. https://doi.org/10.1038/nature08662
23. Ferrell TL, Echenique PM (1985) Phys Rev Lett 55:1526. https://doi.org/10.1103/PhysRevLett.55.1526
24. Garcia-Molina R, Gras-Marti A, Ritchie RH (1985) Phys Rev B 31:121. https://doi.org/10.1103/PhysRevB.31.121
25. Rivacoba A, Zabala N, Echenique PM (1992) Phys Rev Lett 69:3362. https://doi.org/10.1103/PhysRevLett.69.3362
26. Zabala N, Rivacoba A, Echenique PM (1989) Surf Sci 209:465. https://doi.org/10.1016/0039-6028(89)90089-7
27. Echenique PM, Bausells J, Rivacoba A (1987) Phys Rev B 35:1521. https://doi.org/10.1103/PhysRevB.35.1521
28. Ritchie RH, Howie A (1988) Phil Mag A 58:753. https://doi.org/10.1080/01418618808209951
29. Kröger E (1970) Z Phys 235:403. https://doi.org/10.1007/BF01394931
30. García de Abajo FJ (1999) Phys Rev B 59:3095. https://doi.org/10.1103/PhysRevB.59.3095
31. Hage FS, Radtke G, Kepaptsoglou DM, Lazzeri M, Ramasse QM (2020) Science 367:1124. https://doi.org/10.1126/science.aba1136
32. Yankovich AB, Munkhbat B, Baranov DG, Cuadra J, Olsén E, Lourenço-Martins H, Tizei LHG, Kociak M, Olsson E, Shegai T (2019) Nano Lett 19:8171. https://doi.org/10.1021/acs.nanolett.9b03534
33. Pogorzelski R, Yeh C (1973) Phys Rev A 8:137. https://doi.org/10.1103/PhysRevA.8.137
34. Ginzburg VL (1996) Phys Usp 39:973. https://doi.org/10.1070/PU1996v039n10ABEH000171
35. Yamamoto N, Sugiyama H, Toda A (1996) Proc R Soc London Ser A 452:2279. https://doi.org/10.1098/rspa.1996.0122
36. Lucas AA, Šunjić M (1972) Prog Surf Sci 2:75. https://doi.org/10.1016/0079-6816(72)90002-0
37. Olver FWJ, Daalhuis ABO, Lozier DW, Schneider BI, Boisvert RF, Clark CW, Mille BVSBR, Cohl HS, McClain MA (2023) NIST digital library of mathematical functions. Release 1.1.9 of 2023-03-15. https://dlmf.nist.gov/
38. Batson PE, Reyes-Coronado A, Barrera RG, Rivacoba A, Echenique PM, Aizpurua J (2011) Nano Lett 11:3388. https://doi.org/10.1021/nl201795u
39. Christensen T, Yan W, Raza S, Jauho A-P, Mortensen NA, Wubs M (2014) ACS Nano 8:1745. https://doi.org/10.1021/nn406153k
40. Zouros GP, Kolezas GD, Mortensen NA, Tserkezis C (2020) Phys Rev B 101:085416. https://doi.org/10.1103/PhysRevB.101.085416

41. Gonçalves PAD, García de Abajo FJ (2023) Nano Lett 23:4242. https://doi.org/10.1021/acs.nanolett.3c00298
42. Moeferdt M (2017) Nonlocal and nonlinear properties of plasmonic nanostructures within the hydrodynamic Drude model. PhD thesis, Humboldt Universität zu Berlin (Germany). https://edoc.hu-berlin.de/bitstream/handle/18452/18800/moeferdt.pdf?sequence=4
43. Matsukata T, García de Abajo FJ, Sannomiya T (2021) ACS Nano 15:2219. https://doi.org/10.1021/acsnano.0c05624
44. Maciel-Escudero C, Yankovich AB, Munkhbat B, Baranov DG, Hillenbrand R, Olsson E, Aizpurua J, Shegai TO (2023) Nat Commun 14:8478. https://doi.org/10.1038/s41467-023-43813-y
45. Green MA (2008) Sol Energy Mater Sol Cells 92:1305. https://doi.org/10.1016/j.solmat.2008.06.009
46. Frank IM, Tamm IE (1937) Dokl Akad Nauk SSSR 14:109. https://doi.org/10.3367/UFNr.0093.196710o.0388
47. Ginzburg VL, Frank IM (1945) J Phys USSR 9:353
48. Goldsmith P, Jelley JV (1959) Philos Mag 4:836. https://doi.org/10.1080/14786435908238241
49. van de Groep J, Polman A (2013) Opt Express 21:26285. https://doi.org/10.1364/OE.21.026285
50. Fiedler S, Raza S, Ai R, Wang J, Busch K, Stenger N, Mortensen NA, Wolff C (2020) Opt Express 28:13938. https://doi.org/10.1364/OE.393056

Chapter 6
Summary and Conclusions

Abstract This chapter briefly summarize the content of each chapter, highlights the key findings, and the conclusions derived.

Throughout this thesis we investigated several aspects regarding light–matter interaction involving spherical metallic or dielectric nanostructures. A large part of this work was devoted to developing analytic methods, based on the well-established Mie theory, extended to encompass the relevant physics in diverse scenarios, including different excitation sources, optical properties or length scales, in classical and quantum-informed frameworks. In closing, we briefly summarize the content of each chapter, highlight the key findings, and outline the conclusions derived.

Chapter 2 | **Theory**

The main objective of this chapter was to establish a fundamental theoretical description of light and matter, as well as the interactions occurring between them. In this regard, we investigated the optical excitations emerging upon the illumination of dielectric and metallic NPs, namely Mie resonances and LSPs, respectively. Subsequently, we introduced the concepts of strong and weak coupling of such modes with QEs, and derived the relevant physical quantities. Finally, we presented classical and semiclassical analytic and numerical tools used in various parts of the thesis, thereby laying the foundation for the EM problems addressed in the forthcoming chapters.

Chapter 3 | **Strong Coupling in Photonics**

Chapter 3 addressed the strong coupling between optical transitions and EM modes. As the focal point of our investigation, we examined core–shell configurations that combine metallic or dielectric with excitonic components. One of our main interests in this work is the exploration of light–matter interactions at the extreme nanoscale, where nonlocal and quantum effects are expected to manifest. In this spirit, the question we sought to answer was whether the classically predicted coupling in plexcitonic systems deviates significantly from the semiclassical picture presented by various nonlocal schemes of varying complexity. Remarkably, the increased damping did not inhibit the onset of strong coupling, nor did it bring about any substantial changes to the observed coupling strength. Equipped with this knowledge, we then explored whether the chiral properties of matter can be reconfigured in the strong-coupling regime, by examining two configurations: a NP composed of a dielectric

P. E. Stamatopoulou, *Strong Light-Matter Interactions in Extreme Plasmonic and Mie-Resonant Systems*, Springer Theses, https://doi.org/10.1007/978-3-031-67907-0_6

core embedded in an excitonic shell, and a helix consisting of nanometric metallic spheres encapsulated in an excitonic matrix. Indeed, the chiroptical response of both platforms featured the signature of strong coupling, namely, modal anticrossings. This outcome can serve a twofold function, as it can prove beneficial not only in the context of chirality sensing, but also in the characterization of light–matter coupling.

Chapter 4 | **Weak Coupling of Nanoparticles with Quantum Emitters**

This chapter explored how the excitation and emission properties of quantum emitters are altered by nearby plasmonic or Mie-resonant systems. Having identified an optimal orientation and position of the emitter with respect to the resonant structure for maximal fluorescence, we proceeded to assess the enhancement factors achieved with metals compared to dielectrics. The takeaway of this comparison can be summarized to the observation that plasmonic NPs achieve superior enhancement factors, albeit under specific configurations, whereas dielectric ones exhibit greater versatility in this regard, owing to their highly directional and readily controllable scattering of the incident light. Since smaller NPs achieve higher field localization and enhancement, it is rather imperative to investigate whether such properties survive at the nanoscale, when nonlocality is factored in. In this interplay, the localized nature of the emitter exposed the inadequacies of local descriptions, which were shown to considerably overestimate the excitation and fluorescence enhancements achievable. Understanding both the capabilities and constraints associated with nanoantennas of diverse materials and sizes can contribute to the design of more efficient sensing platforms.

Chapter 5 | **Interaction of Nanoparticles with Electron Beams**

In Chap. 5 we examined the optical response of plasmonic and dielectric NPs excited by swift electrons. To this end, we devised an analytic method for calculating the energy loss undergone by the electrons as they traverse a spherical scatterer, along with the radiation emitted in the process. As a first step, we juxtaposed our results with those acquired with a numerical method, thereby demonstrating the respective strengths and limitations of each approach. Certainly, the analytic approach is more advantageous in retrieving the mechanisms that underlie the spectral features observed. Here, we directed our focus on penetrating electrons, as they potentially capture more physical processes than aloof ones, e.g. bulk plasmons in metallic NPs. Furthermore, our analysis of dielectric NPs identified specific energy ranges where other sources of energy loss, beyond the Mie resonances, come to the fore, including transition and Cherenkov radiation, and interband transitions. These findings led us to the conclusion that, even though penetrating electron beams unveil a richer spectrum of physical phenomena, disentangling the origin of the respective observations can be a challenging task. In light of the remarkable progress of electron-beam spectroscopy, our analysis could play a vital role in interpreting the plethora of experimental data.

We believe that there are several open questions arising from this thesis that could inspire and lay the theoretical groundwork for future research. Firstly, chiral polaritons are very promissing for enhanced CD sensing. As a next step to the work presented here one could explore how resonant chiral cavities interact with chiral excitonic materials, with an aim to enhance the CD of solutions of chiral molecules through the incorporation of chiral NPs. Continuing in the context of NP-enhanced sensing, nanometric plasmonic NPs can mediate the access to high-order multipolar transitions which, of course, requires a rigorous study in the scope of mesoscopic plasmonics.

Regarding our work on electron-beam spectroscopy, a next natural step would be to apply the analytic method developed for CL and EELS to perform a systematic study of the various mechanisms, such as Cherenkov and transition radiation, coexisting and interfering with the EM modes of the NPs. Secondly, since the method is based on the framework of Mie theory, it can be extended with relative ease to account for nonlocal effects, by introducing the appropriately modified Mie coefficients. In that regard, it would be particularly interesting to explore quantum effects in EELS with penetrating electron beams, and juxtapose the results with experiments. Such a comparison could shed light on the significance of bulk quantum effects, which cannot be captured by surface response functions. Another logical progression would be to probe strong coupling between dielectric or metallic NPs and excitons with electron beams, and examine whether and how the coupling strength is affected by the beam velocity and the impact parameter. Finally, we speculate that the majority of the methods developed concerning spherical cavities can be extended to cylindrical or elliptical geometries via the Extended Boundary Condition Method [1].

In the evolving landscape of nanophotonics, which aims for ever stronger light–matter interactions at progressively smaller scales, our research, though not exhaustive, aspires to remain relevant. In pursuit of this quest, we developed the theoretical tools that will prove practical in refining our fundamental understanding, interpreting experimental observations, and designing novel photonic platforms. With the final chapter written, we hope that our effort has pushed the boundaries of what is known just a little further.

Reference

1. Mishchenko MI, Travis LD, Lacis AA (2002) Scattering, absorption, and emission of light by small particles. Cambridge University Press

Appendix A
The Helmholtz Equation in Spherical Coordinates

Most problems addressed in this thesis boil down to finding the solution to the scalar Helmholtz equation [1]

$$\left(\nabla^2 + k^2\right)\psi = 0\,. \tag{A.1}$$

For systems with spherical symmetry, it is most convenient to solve the equation in spherical coordinates (r, θ, ϕ)

$$\left[\frac{1}{r^2}\frac{\partial}{\partial r}\left(r^2\frac{\partial}{\partial r}\right) + \frac{1}{r^2\sin\theta}\frac{\partial}{\partial\theta}\left(\sin\theta\frac{\partial}{\partial\theta}\right) + \frac{1}{r^2\sin^2\theta}\frac{\partial^2}{\partial\phi^2} + k^2\right]\psi(r,\theta,\phi) = 0\,. \tag{A.2}$$

One can introduce in Eq. (A.2) the orbital angular momentum operator $\mathbf{L}$, defined as [1]

$$\mathbf{L} = -i(\mathbf{r}\times\nabla) = i\left(\hat{\boldsymbol{\theta}}\frac{1}{\sin\theta}\frac{\partial}{\partial\phi} - \hat{\boldsymbol{\phi}}\frac{\partial}{\partial\theta}\right), \tag{A.3a}$$

with

$$\mathrm{L}^2 = -\frac{1}{\sin\theta}\frac{\partial}{\partial\theta}\left(\sin\theta\frac{\partial}{\partial\theta}\right) - \frac{1}{\sin^2\theta}\frac{\partial^2}{\partial\phi^2}\,. \tag{A.3b}$$

Then, separation of variables to $\psi(r,\theta,\phi) = R(r)Y(\theta,\phi)$ gives

$$\left[\frac{1}{r^2}\frac{\partial}{\partial r}\left(r^2\frac{\partial}{\partial r}\right) - \frac{\mathrm{L}^2}{r^2} + k^2\right]R(r)Y(\theta,\phi) = 0 \tag{A.4}$$

and allows to separate the radial from the angular part of the equation, as

$$\left[\frac{d^2}{dr^2} + \frac{2}{r}\frac{d}{dr} + k^2 - \frac{\ell(\ell+1)}{r^2}\right]R(r) = 0\,, \tag{A.5a}$$

$$\mathrm{L}^2 Y(\theta,\phi) = \ell(\ell+1)Y(\theta,\phi)\,. \tag{A.5b}$$

P. E. Stamatopoulou, *Strong Light-Matter Interactions in Extreme Plasmonic and Mie-Resonant Systems*, Springer Theses, https://doi.org/10.1007/978-3-031-67907-0

Equation (A.5b) can be further split into its polar and azimuthal angle dependence $Y(\theta, \phi) = \Theta(\theta)\Phi(\phi)$, yielding two angular differential equations:

$$\left[\frac{d^2}{d\theta^2} + \frac{\cos\theta}{\sin\theta}\frac{d}{d\theta} + \ell(\ell+1) - \frac{m^2}{\sin^2\theta}\right]\Theta(\theta) = 0\,, \tag{A.6a}$$

$$\left[\frac{d^2}{d\phi^2} - m^2\right]\Phi(\phi) = 0\,. \tag{A.6b}$$

One can easily show that the solutions to the angular equation (A.5b) are the spherical harmonics Y_ℓ^m [1]. The degree ℓ and order m, that emerged as separation variables, are required by the boundary conditions to be integers, and correspond to the orbital angular momentum and the projection of the orbital angular momentum quantum numbers, respectively. The individual solutions of the radial and the angular part are discussed in more detail in the following appendices. In Appendix B we discuss the spherical Bessel functions, namely the solutions to the radial differential equation, whereas in Appendix C we provide a detailed account of the scalar and vector spherical harmonics, the latter arising as solutions to the angular part of the vector Helmholtz equation.

Appendix B
Bessel Functions

B.1 Spherical Bessel Functions

Definition The family of spherical Bessel functions is obtained as a solution to the radial part of the Helmholtz equation when solved in spherical coordinates. Assuming a general solution of the form $\psi(r, \theta, \phi) = R(r)\Theta(\theta)\Phi(\phi)$, the radial differential equation takes the form [see Eq. (A.5a)]

$$r^2 \frac{d^2 R}{dr^2} + 2r \frac{dR}{dr} + \left[k^2 r^2 - \ell(\ell + 1)\right] R = 0 \,, \tag{B.1}$$

where k is a parameter originating from the Helmholtz equation, and $\ell(\ell + 1)$ is the separation constant associated with its angular part, as seen in Eqs. (A.4) and (A.5b) respectively. Equation (B.1) is equivalent to

$$r^2 \frac{d^2 Z}{dr^2} + r \frac{dZ}{dr} + \left[k^2 r^2 - \left(\ell + \frac{1}{2}\right)\right] Z = 0 \,, \tag{B.2}$$

where we set $R(kr) = Z(kr)/\sqrt{kr}$. This is the Bessel equation of order $\ell + 1/2$, admitting two linearly independent solutions: the Bessel function (of the first kind) J_n, and the Neumann function Y_n (also called Bessel function of the second kind) [1]. Therefore, R is a linear combination of spherical Bessel and Neumann functions, defined as

$$j_\ell(z) = \sqrt{\frac{\pi}{2z}} J_{\ell+\frac{1}{2}}(z), \tag{B.3a}$$

$$y_\ell(z) = \sqrt{\frac{\pi}{2z}} Y_{\ell+\frac{1}{2}}(z). \tag{B.3b}$$

P. E. Stamatopoulou, *Strong Light-Matter Interactions in Extreme Plasmonic and Mie-Resonant Systems*, Springer Theses, https://doi.org/10.1007/978-3-031-67907-0

We may define the Hankel functions of the first, h_ℓ^+, and second kind, h_ℓ^-, as

$$h_\ell^\pm(z) = j_\ell(z) \pm i y_\ell(z) , \tag{B.4}$$

which are also solutions to Eq. (B.1), being a linear combination of j_ℓ and y_ℓ. It is then clear that

$$j_\ell(z) = \frac{1}{2}\left[h_\ell^+(z) + h_\ell^-(z)\right] . \tag{B.5}$$

Power Series The spherical Bessel and Neumann functions can be evaluated via the power series [2]

$$j_\ell(z) = z^\ell \sum_{n=0}^{\infty} \frac{(-z^2/2)^n}{n!(2\ell+2n+1)!!} , \tag{B.6a}$$

$$y_\ell(z) = -\frac{1}{z^{\ell+1}} \sum_{n=0}^{\ell} \frac{(2\ell-2n-1)!!(z^2/2)^n}{n!} + \frac{(-1)^{\ell+1}}{z^{\ell+1}} \sum_{n=\ell+1}^{\infty} \frac{(-z^2/2)^n}{n!(2n-2\ell-1)!!} , \tag{B.6b}$$

where !! stands for the double factorial.

Limiting Forms For small arguments $z \to 0$, the spherical Bessel functions take the form

$$j_\ell(z) \sim \frac{z^\ell}{(2\ell+1)!!} , \tag{B.7a}$$

$$y_\ell(z), -ih_\ell^+(z), ih_\ell^-(z) \sim -\frac{(2\ell-1)!!}{z^{\ell+1}} , \tag{B.7b}$$

while for large arguments $z \to \infty$ they follow the asymptotic expressions

$$j_\ell(z) \sim \frac{1}{z}\sin\left(z - \frac{\ell\pi}{2}\right) , \tag{B.8a}$$

$$y_\ell(z) \sim -\frac{1}{z}\cos\left(z - \frac{\ell\pi}{2}\right) , \tag{B.8b}$$

$$h_\ell^\pm(z) \sim (\mp i)^{\ell+1}\frac{e^{\pm iz}}{z} . \tag{B.8c}$$

Wronskian Determinant In our calculations we frequently encounter the Wronskian determinant

$$\mathscr{W}\{j_\ell(z), y_\ell(z)\} = \begin{vmatrix} j_\ell(z) & y_\ell(z) \\ j_\ell'(z) & y_\ell'(z) \end{vmatrix} = \frac{1}{z^2} , \tag{B.9}$$

where the prime denotes the derivative of the function with respect to its argument. By introducing relation (B.4) we obtain

$$\mathscr{W}\{j_\ell(z), h_\ell^\pm(z)\} = \pm\frac{i}{z^2}\,, \tag{B.10a}$$

$$\mathscr{W}\{h_\ell^+(z), h_\ell^-(z)\} = -\frac{2i}{z^2}\,. \tag{B.10b}$$

Recurrence Relations All spherical Bessel-type functions $f_\ell(z)$ follow the same recurrence relations

$$f_\ell(z) = \frac{z}{2\ell+1}\left[f_{\ell-1}(z) + f_{\ell+1}(z)\right]\,, \tag{B.11a}$$

$$f_\ell'(z) = \frac{1}{2\ell+1}\left[\ell f_{\ell-1}(z) - (\ell+1) f_{\ell+1}(z)\right]\,, \tag{B.11b}$$

$$f_\ell'(z) = f_{\ell-1}(z) - \frac{\ell+1}{z} f_\ell(z)\,, \tag{B.11c}$$

$$f_\ell'(z) = -f_{\ell+1}(z) + \frac{\ell}{z} f_\ell(z)\,. \tag{B.11d}$$

B.2 Modified Bessel Functions

Definition The modified Bessel functions $I_m(z)$ and $K_m(z)$ are solutions to the modified Bessel equation [1]

$$z^2\frac{d^2Z}{dz^2} + z\frac{dZ}{dz} - (z^2+m^2)Z = 0\,, \tag{B.12}$$

obtained from the usual Bessel equation (B.2) by replacing the argument z with iz.

Connection Relations I_m is the modified Bessel function of the first kind, related to the conventional Bessel function via

$$I_m(z) = i^{-m} J_m(iz)\,. \tag{B.13}$$

K_m is the modified Bessel function of the second kind, defined as

$$K_m(z) = \frac{\pi}{2} i^{m+1}\left[J_m(iz) + iY_m(iz)\right]\,. \tag{B.14}$$

Power Series Function I_m is given by the power series

$$I_m(z) = \left(\frac{z}{2}\right)^m \sum_{n=0}^{\infty} \frac{(z^2/4)^n}{n!\Gamma(m+n+1)}\,, \tag{B.15}$$

where the gamma Γ function of an integer n is equal to $\Gamma(n) = (n-1)!$.

Wronskian Determinant The Wronskian determinant of the modified Bessel functions is given by

$$\mathscr{W}\{I_m(z), K_m(z)\} = -\frac{1}{z} . \tag{B.16}$$

Recurrence Relations The modified Bessel functions satisfy the following recurrence relations

$$I_m(z) = \frac{z}{2m}[I_{m-1}(z) - I_{m+1}(z)] , \tag{B.17a}$$

$$I'_m(z) = \frac{1}{2}[I_{m-1}(z) + I_{m+1}(z)] , \tag{B.17b}$$

$$I'_m(z) = I_{m\pm 1}(z) \pm \frac{m}{z} I_m(z) , \tag{B.17c}$$

$$K_m(z) = -\frac{z}{2m}[K_{m-1}(z) - K_{m+1}(z)] , \tag{B.17d}$$

$$K'_m(z) = -\frac{1}{2}[K_{m-1}(z) + K_{m+1}(z)] , \tag{B.17e}$$

$$K'_m(z) = -K_{m\pm 1}(z) \pm \frac{m}{z} K_m(z) . \tag{B.17f}$$

For negative order $-m$ we have

$$I_{-m}(z) = I_m(z) , \tag{B.18a}$$

$$K_{-m}(z) = K_m(z) . \tag{B.18b}$$

Graf's Addition Theorem Given a distance $r = \left[\rho^2 + \rho'^2 - 2\rho\rho' \cos(\phi - \phi')\right]^{1/2}$ in polar coordinates, an explicit version of Graf's addition theorem for K_m is given by [3][1]

$$K_0(\eta r) = \sum_{m=-\infty}^{\infty} K_m(\eta\rho_>) I_m(\eta\rho_<) \mathrm{e}^{im(\phi-\phi')} , \tag{B.19}$$

where $\rho_{</>}$ stands for the smaller/greater between ρ and ρ'.

[1] Can be derived from Eq. 9.1.79 (pp. 363) in Ref. [3] or from Eq. 10.23.7 in Ref. [2].

Appendix C
Scalar and Vector Spherical Harmonics

C.1 Scalar Spherical Harmonics

Definition The scalar spherical harmonics, or simply spherical harmonics $Y_\ell^m(\hat{\mathbf{r}})$ constitute the angular part of the solution to the Helmholtz equation in spherical coordinates [see Eq. (A.4)], where $\hat{\mathbf{r}}$ stands for the dependence on the solid angle (θ, ϕ). They form a complete set of orthogonal functions that satisfy

$$\nabla^2 Y_\ell^m(\hat{\mathbf{r}}) = -\frac{\mathrm{L}^2}{r^2} Y_\ell^m(\hat{\mathbf{r}}) = -\frac{\ell(\ell+1)}{r^2} Y_\ell^m(\hat{\mathbf{r}}) \,. \tag{C.1}$$

They are eigenfunctions of the operators L^2 and $\mathrm{L}_z = \mathbf{L} \cdot \hat{\mathbf{z}}$ with $\mathbf{L}$ defined as in Eq. (A.3a), and they satisfy the following relations

$$\mathrm{L}^2 Y_\ell^m(\hat{\mathbf{r}}) = \ell(\ell+1) Y_\ell^m(\hat{\mathbf{r}}) \,, \tag{C.2}$$

$$\mathrm{L}_z Y_\ell^m(\hat{\mathbf{r}}) = m Y_\ell^m(\hat{\mathbf{r}}) \,, \tag{C.3}$$

$$\mathrm{L}_\pm Y_\ell^m(\hat{\mathbf{r}}) = 2\alpha_\ell^{\pm m} Y_\ell^{m\pm 1}(\hat{\mathbf{r}}) \,, \tag{C.4}$$

where $\mathrm{L}_\pm = \mathrm{L}_x \pm i\mathrm{L}_y$ are the ladder operators, and

$$\alpha_\ell^m = \frac{1}{2}\sqrt{(\ell - m)(\ell + m + 1)} \,. \tag{C.5}$$

Their values can be found by the formula [1]

$$Y_\ell^m(\hat{\mathbf{r}}) = (-1)^m \sqrt{\frac{2\ell+1}{4\pi}\frac{(\ell-m)!}{(\ell+m)!}} P_\ell^m(\cos\theta) \mathrm{e}^{im\phi} \,, \tag{C.6}$$

P. E. Stamatopoulou, *Strong Light-Matter Interactions in Extreme Plasmonic and Mie-Resonant Systems*, Springer Theses, https://doi.org/10.1007/978-3-031-67907-0

where the degree ℓ is the quantum number associated with the orbital angular momentum and assumes values $\ell = 0, 1, \ldots,$ and the order m is related to the projection of the orbital angular momentum, with $m = -\ell, -\ell+1, \ldots, \ell-1, \ell$. The associated Legendre polynomials P_ℓ^m are defined for $m \geq 0$ as

$$P_\ell^m(x) = \frac{1}{2^\ell \ell!}\left(1-x^2\right)^{m/2} \frac{d^{\ell+m}}{dx^{\ell+m}}\left(x^2-1\right)^\ell , \quad x = \cos\theta \tag{C.7}$$

while for $m < 0$, on can make use of the following relation

$$P_\ell^{-m}(x) = (-1)^m \frac{(\ell-m)!}{(\ell+m)!} P_\ell^m(x) . \tag{C.8}$$

Using Eq. (C.8) we find

$$Y_\ell^{m*}(\hat{\mathbf{r}}) = (-1)^m Y_\ell^{-m}(\hat{\mathbf{r}}) . \tag{C.9}$$

To avoid confusion with other conventions found in literature, we emphasize that in the notation used here the Condon-Shortley phase $(-1)^m$ is explicitly contained in the definition of the spherical harmonics (C.6), rather than inside the associated Legendre polynomials (C.7).

Orthonormality and Completeness The spherical harmonics are orthonormal

$$\int_{\theta=0}^{\pi}\int_{\phi=0}^{2\pi} d\Omega\, Y_\ell^m(\hat{\mathbf{r}}) Y_{\ell'}^{m'*}(\hat{\mathbf{r}}) = \delta_{\ell\ell'}\delta_{mm'} , \tag{C.10}$$

where δ_{ij} is the Kronecker delta and $d\Omega = \sin\theta d\theta d\phi$ the infinitesimal solid angle. Given two vectors $\mathbf{r}$ and $\mathbf{r}'$, the spherical harmonics satisfy the completeness relation

$$\sum_{\ell=0}^{\infty}\sum_{m=-\ell}^{\ell} Y_\ell^m(\hat{\mathbf{r}}) Y_\ell^{m*}(\hat{\mathbf{r}}') = \delta(\hat{\mathbf{r}} - \hat{\mathbf{r}}') , \tag{C.11}$$

with $\delta(x)$ being the Dirac delta function.

Parity The spherical harmonics have definite parity $Y_\ell^m(-\hat{\mathbf{r}}) = (-1)^\ell Y_\ell^m(\hat{\mathbf{r}})$, or, in terms of angles

$$Y_\ell^m(\pi-\theta, \pi+\phi) = (-1)^\ell Y_\ell^m(\theta, \phi) . \tag{C.12}$$

Addition Theorem Given two vectors $\mathbf{r}$ and $\mathbf{r}'$ with relative angle γ given by

$$\cos\gamma = \cos\theta\cos\theta' + \sin\theta\sin\theta'\cos(\phi - \phi') , \tag{C.13}$$

the addition theorem states that

$$P_\ell(\cos\gamma) = \frac{4\pi}{2\ell+1}\sum_{m=-\ell}^{\ell} Y_\ell^m(\hat{\mathbf{r}}) Y_\ell^{m*}(\hat{\mathbf{r}}') . \tag{C.14}$$

When $\mathbf{r} = \mathbf{r}'$ this leads to

$$\sum_{m=-\ell}^{\ell} |Y_\ell^m(\hat{\mathbf{r}})|^2 = \frac{2\ell+1}{4\pi} \,. \tag{C.15}$$

Special Case We mention as a useful special case the relation

$$Y_\ell^m(0,\phi) = \sqrt{\frac{2\ell+1}{4\pi}}\,\delta_{m0} \,. \tag{C.16}$$

Recurrence Relations The following relations connect spherical harmonics of different degrees and orders

$$\cos\theta Y_\ell^m(\hat{\mathbf{r}}) = \zeta_{\ell+1}^m Y_{\ell+1}^m(\hat{\mathbf{r}}) + \zeta_\ell^m Y_{\ell-1}^m(\hat{\mathbf{r}}) \,, \tag{C.17a}$$

$$\mathrm{e}^{\pm i\phi}\sin\theta Y_\ell^m(\hat{\mathbf{r}}) = \pm\big[\gamma_\ell^{\mp m} Y_{\ell-1}^{m\pm1}(\hat{\mathbf{r}}) - \gamma_{\ell+1}^{\pm m+1} Y_{\ell+1}^{m\pm1}(\hat{\mathbf{r}})\big] \,, \tag{C.17b}$$

$$m\cot\theta Y_\ell^m(\hat{\mathbf{r}}) = -\big[\alpha_\ell^m \mathrm{e}^{-i\phi} Y_\ell^{m+1}(\hat{\mathbf{r}}) + \alpha_\ell^{-m}\mathrm{e}^{i\phi} Y_\ell^{m-1}(\hat{\mathbf{r}})\big] \,, \tag{C.17c}$$

where

$$\zeta_\ell^m = \sqrt{\frac{(\ell+m)(\ell-m)}{(2\ell-1)(2\ell+1)}} \,, \tag{C.18}$$

$$\gamma_\ell^m = \sqrt{\frac{(\ell+m)(\ell+m-1)}{(2\ell-1)(2\ell+1)}} \,. \tag{C.19}$$

Other Recurrent Relations In Appendix (F) we frequently encounter the partial derivative ladder operator $\partial_\pm = \partial_x \pm i\partial_y$, acting on the product $f_\ell Y_\ell^m$, where f_ℓ stands for any spherical Bessel function. In cylindrical coordinates (ρ, ϕ, z)

$$\partial_\pm = \mathrm{e}^{\pm i\phi}\left\{\frac{\partial}{\partial\rho} \pm \frac{i}{\rho}\frac{\partial}{\partial\phi}\right\} . \tag{C.20}$$

Since $r = \sqrt{\rho^2+z^2}$ and $\theta = \arccos(z/r)$, we have

$$\begin{aligned}\partial_\pm\{f_\ell(kr)Y_\ell^m(\hat{\mathbf{r}})\} = \mathrm{e}^{\pm i\phi}\Bigg\{&\frac{zf_\ell\left(k\sqrt{\rho^2+z^2}\right)}{\rho^2+z^2}\left[\alpha_\ell^m\mathrm{e}^{-i\phi}Y_\ell^{m+1}(\hat{\mathbf{r}}) - \alpha_\ell^{-m}\mathrm{e}^{i\phi}Y_\ell^{m-1}(\hat{\mathbf{r}})\right]\\ &+\frac{k\rho f_\ell'\left(k\sqrt{\rho^2+z^2}\right)}{\sqrt{\rho^2+z^2}}Y_\ell^m(\hat{\mathbf{r}}) \mp \frac{m}{\rho}f_\ell\left(k\sqrt{\rho^2+z^2}\right)Y_\ell^m(\hat{\mathbf{r}})\Bigg\} ,\end{aligned} \tag{C.21}$$

C.2 Vector Spherical Harmonics

Definition Adopting the notation used in Ref. [4], the vector spherical harmonics are defined through

$$\mathbf{X}_{\ell m}(\hat{\mathbf{r}}) = \frac{1}{s_\ell}\mathbf{L}Y_\ell^m(\hat{\mathbf{r}}) , \tag{C.22}$$

where $s_\ell = \sqrt{\ell(\ell+1)}$. Substitution of the angular momentum operator (A.3a) in Eq. (C.22) gives

$$\mathbf{X}_{\ell m}(\hat{\mathbf{r}}) = X_{\ell m,\theta}(\hat{\mathbf{r}})\hat{\boldsymbol{\theta}} + X_{\ell m,\phi}(\hat{\mathbf{r}})\hat{\boldsymbol{\phi}} , \tag{C.23}$$

with

$$X_{\ell m,\theta}(\hat{\mathbf{r}}) = \frac{i}{s_\ell \sin\theta}\frac{\partial Y_\ell^m(\hat{\mathbf{r}})}{\partial\phi} = -\frac{m}{s_\ell \sin\theta}Y_\ell^m(\hat{\mathbf{r}}) , \tag{C.24a}$$

$$X_{\ell m,\phi}(\hat{\mathbf{r}}) = -\frac{i}{s_\ell}\frac{\partial Y_\ell^m(\hat{\mathbf{r}})}{\partial\theta} = -\frac{i}{s_\ell}\left[\alpha_\ell^m \mathrm{e}^{-i\phi}Y_\ell^{m+1}(\hat{\mathbf{r}}) - \alpha_\ell^{-m}\mathrm{e}^{i\phi}Y_\ell^{m-1}(\hat{\mathbf{r}})\right] . \tag{C.24b}$$

By definition $\mathbf{X}_{00}(\hat{\mathbf{r}}) = 0$, while for $\ell \geq 1$ we have

$$\begin{aligned} s_\ell\mathbf{X}_{\ell m}(\hat{\mathbf{r}}) &= \left[\alpha_\ell^{-m}\cos\theta \mathrm{e}^{i\phi}Y_\ell^{m-1}(\hat{\mathbf{r}}) - m\sin\theta Y_\ell^m(\hat{\mathbf{r}}) + \alpha_\ell^m\cos\theta\mathrm{e}^{-i\phi}Y_\ell^{m+1}(\hat{\mathbf{r}})\right]\hat{\boldsymbol{\theta}} \\ &\quad + i\left[\alpha_\ell^{-m}\mathrm{e}^{i\phi}Y_\ell^{m-1}(\hat{\mathbf{r}}) - \alpha_\ell^m\mathrm{e}^{-i\phi}Y_\ell^{m+1}(\hat{\mathbf{r}})\right]\hat{\boldsymbol{\phi}} \qquad \text{(C.25a)} \\ &= \alpha_\ell^{-m}Y_\ell^{m-1}(\hat{\mathbf{r}})(\hat{\mathbf{x}}+i\hat{\mathbf{y}}) + mY_\ell^m(\hat{\mathbf{r}})\hat{\mathbf{z}} + \alpha_\ell^m Y_\ell^{m+1}(\hat{\mathbf{r}})(\hat{\mathbf{x}}-i\hat{\mathbf{y}}) . \qquad \text{(C.25b)} \end{aligned}$$

From Eq. (C.22) it follows that

$$\mathbf{L}\cdot\mathbf{X}_{\ell m}(\hat{\mathbf{r}}) = s_\ell Y_\ell^m(\hat{\mathbf{r}}) , \tag{C.26}$$

$$\mathrm{L}^2\mathbf{X}_{\ell m}(\hat{\mathbf{r}}) = \ell(\ell+1)\mathbf{X}_{\ell m}(\hat{\mathbf{r}}) , \tag{C.27}$$

$$\mathbf{X}_{\ell m}^*(\hat{\mathbf{r}}) = \frac{1}{s_\ell}\left[-i\mathbf{r}\times\nabla Y_\ell^m(\hat{\mathbf{r}})\right]^* = (-1)^{m+1}\mathbf{X}_{\ell\,-m}(\hat{\mathbf{r}}) , \tag{C.28}$$

and

$$\mathbf{X}_{\ell m}(\hat{\mathbf{r}})\times\hat{\mathbf{r}} = X_{\ell m,\phi}(\hat{\mathbf{r}})\hat{\boldsymbol{\theta}} - X_{\ell m,\theta}(\hat{\mathbf{r}})\hat{\boldsymbol{\phi}} = -\frac{ir}{s_\ell}\nabla Y_\ell^m(\hat{\mathbf{r}}) , \tag{C.29}$$

while

$$\hat{\mathbf{r}}\times\left[\mathbf{X}_{\ell m}(\hat{\mathbf{r}})\times\hat{\mathbf{r}}\right] = \mathbf{X}_{\ell m}(\hat{\mathbf{r}}) . \tag{C.30}$$

Orthonormality The vector spherical harmonics comply with the orthonormality relations

$$\int d\Omega\, \mathbf{X}^*_{\ell' m'}(\hat{\mathbf{r}}) \cdot \mathbf{X}_{\ell m}(\hat{\mathbf{r}}) = \delta_{\ell\ell'}\delta_{mm'} \,, \tag{C.31a}$$

$$\int d\Omega\, \left[\hat{\mathbf{r}} \times \mathbf{X}^*_{\ell' m'}(\hat{\mathbf{r}})\right] \cdot \mathbf{X}_{\ell m}(\hat{\mathbf{r}}) = 0 \,. \tag{C.31b}$$

Divergence, Curl and Laplacian Acting with operators $\nabla\cdot$ and $\nabla\times$, we obtain

$$\nabla \cdot \mathbf{X}_{\ell m}(\hat{\mathbf{r}}) = 0 \,, \tag{C.32}$$

and

$$\begin{aligned} \nabla \times \mathbf{X}_{\ell m}(\hat{\mathbf{r}}) &= \frac{1}{r}\left[i s_\ell Y_\ell^m(\hat{\mathbf{r}})\hat{\mathbf{r}} - X_{\ell m,\phi}(\hat{\mathbf{r}})\hat{\boldsymbol{\theta}} + X_{\ell m,\theta}(\hat{\mathbf{r}})\hat{\boldsymbol{\phi}} \right] \\ &= \frac{1}{r}\left[i s_\ell Y_\ell^m(\hat{\mathbf{r}})\hat{\mathbf{r}} - \mathbf{X}_{\ell m}(\hat{\mathbf{r}}) \times \hat{\mathbf{r}} \right] \\ &= \frac{i}{s_\ell}\left[\frac{s_\ell^2}{r} Y_\ell^m(\hat{\mathbf{r}})\hat{\mathbf{r}} + \nabla Y_\ell^m(\hat{\mathbf{r}}) \right] . \end{aligned} \tag{C.33}$$

We can calculate the vector Laplacian ∇^2, using Eq. (C.27)

$$\nabla^2 \mathbf{X}_{\ell m}(\hat{\mathbf{r}}) = \left[\frac{1}{r^2}\frac{\partial}{\partial r}\left(r^2 \frac{\partial}{\partial r} \right) - \frac{\mathrm{L}^2}{r^2} \right] \mathbf{X}_{\ell m}(\hat{\mathbf{r}}) = -\frac{\ell(\ell+1)}{r^2}\mathbf{X}_{\ell m}(\hat{\mathbf{r}}) \,. \tag{C.34}$$

It is useful for Mie-theory calculations to determine the divergence, curl, and Laplacian operator acting on the product $f_\ell(kr)Y_\ell^m(\hat{\mathbf{r}})$, where f_ℓ is any spherical Bessel function (see Appendix B). We find

$$\nabla \cdot \left[f_\ell(kr)\mathbf{X}_{\ell m}(\hat{\mathbf{r}}) \right] = 0 \,, \tag{C.35}$$

$$\nabla \times \left[f_\ell(kr)\mathbf{X}_{\ell m}(\hat{\mathbf{r}}) \right] = k\left\{ i s_\ell \frac{f_\ell(kr)}{kr} Y_\ell^m(\hat{\mathbf{r}})\hat{\mathbf{r}} - \left[f_\ell'(kr) + \frac{f_\ell(kr)}{kr} \right] \mathbf{X}_{\ell m}(\hat{\mathbf{r}}) \times \hat{\mathbf{r}} \right\} \tag{C.36}$$

$$= i s_\ell \frac{f_\ell(kr)}{r} Y_\ell^m(\hat{\mathbf{r}})\hat{\mathbf{r}} - \frac{1}{r}\frac{\partial}{\partial r}\left[r f_\ell(kr) \right] \mathbf{X}_{\ell m}(\hat{\mathbf{r}}) \times \hat{\mathbf{r}} \,, \tag{C.37}$$

and

$$\begin{aligned} \nabla^2 \left[f_\ell(kr)\mathbf{X}_{\ell m}(\hat{\mathbf{r}}) \right] &= \left[\frac{1}{r^2}\frac{\partial}{\partial r}\left(r^2 \frac{\partial}{\partial r} \right) - \frac{\mathrm{L}^2}{r^2} \right] f_\ell(kr)\mathbf{X}_{\ell m}(\hat{\mathbf{r}}) \,, \\ &= \frac{1}{r^2}\left[-k^2 r^2 + \ell(\ell+1) - \mathrm{L}^2 \right] f_\ell(kr)\mathbf{X}_{\ell m}(\hat{\mathbf{r}}) \\ &= -k^2 f_\ell(kr)\mathbf{X}_{\ell m}(\hat{\mathbf{r}}) \,, \end{aligned} \tag{C.38}$$

where for the last expression we used Eq. (B.1). Using the properties above one can show that

$$\nabla \cdot \mathbf{L}\big[f_\ell(kr)Y_\ell^m(\hat{\mathbf{r}})\big] = 0, \mathbf{L} \cdot \nabla\big[f_\ell(kr)Y_\ell^m(\hat{\mathbf{r}})\big] = 0 \tag{C.39a}$$

$$\mathbf{L} \cdot (\nabla \times \mathbf{L})\big[f_\ell(kr)Y_\ell^m(\hat{\mathbf{r}})\big] = 0, (\nabla \times \mathbf{L}) \cdot \mathbf{L}\big[f_\ell(kr)Y_\ell^m(\hat{\mathbf{r}})\big] = 0,$$
$$(\nabla \times \mathbf{L}) \cdot \big[f_\ell(kr)\mathbf{X}_{\ell m}(\hat{\mathbf{r}})\big] = 0. \tag{C.39b}$$

Other Recurrent Relations From the orthonormality relations one can easily show that the following expressions, encountered frequently in Mie-theory calculations, hold for any spherical Bessel function f_ℓ and g_ℓ

$$\int d\Omega\big[\mathbf{X}^*_{\ell' m'}(\hat{\mathbf{r}}) \times \mathbf{X}_{\ell m}(\hat{\mathbf{r}})\big] \cdot \hat{\mathbf{r}} = 0\,, \tag{C.40a}$$

$$\int d\Omega\, \mathbf{X}^*_{\ell' m'}(\hat{\mathbf{r}}) \cdot \big[\nabla \times f_\ell(kr)\mathbf{X}_{\ell m}(\hat{\mathbf{r}})\big] = 0\,, \tag{C.40b}$$

$$\int d\Omega\big[\nabla \times f_{\ell'}(kr)\mathbf{X}_{\ell' m'}(\hat{\mathbf{r}})\big]^* \times \big[\nabla \times g_\ell(kr)\mathbf{X}_{\ell m}(\hat{\mathbf{r}})\big] \cdot \hat{\mathbf{r}} = 0\,, \tag{C.40c}$$

$$\int d\Omega\big[\hat{\mathbf{r}} \times \mathbf{X}^*_{\ell' m'}(\hat{\mathbf{r}})\big] \cdot \big[\nabla \times f_\ell(kr)\mathbf{X}_{\ell m}(\hat{\mathbf{r}})\big] = \frac{1}{r}\frac{\partial}{\partial r}\left[r f_\ell(kr)\right] \delta_{\ell\ell'}\delta_{mm'}\,, \tag{C.40d}$$

$$\int d\Omega\big[f_{\ell'}(kr)\mathbf{X}_{\ell' m'}(\hat{\mathbf{r}})\big]^* \times \big[\nabla \times g_\ell(kr)\mathbf{X}_{\ell m}(\hat{\mathbf{r}})\big] \cdot \hat{\mathbf{r}} = f_\ell^*(kr)\frac{1}{r}\frac{\partial}{\partial r}\left[r g_\ell(kr)\right] \delta_{\ell\ell'}\delta_{mm'}\,. \tag{C.40e}$$

Appendix D
Mie Theory

D.1 Simple Isotropic Medium

In the present appendix we provide a detailed account of Mie theory [4, 5]. The problem we intent to solve here is depicted in Fig. 2.3; a spherical NP, composed of a simple isotropic material, is embedded in a non-absorbing host medium, and the system is illuminated by a plane EM wave of wavelength comparable to its size. To study the absorption and scattering of the system one has to first calculate the EM field everywhere in space, i.e. inside (region I) and surrounding the NP (II).

Electromagnetic Fields Any smooth vector field, as the electric field, can be decomposed into a transverse magnetic (labeled M), a transverse electric (labeled E) and a longitudinal component (labeled L), using the dyadic identity [6]

$$\mathbf{E} = \nabla\psi^{\mathrm{L}} + \mathbf{L}\psi^{\mathrm{M}} + \frac{i}{k}\nabla \times \mathbf{L}\psi^{\mathrm{E}} , \tag{D.1}$$

where k is the wavenumber inside an arbitrary medium characterized by optical parameters ε, μ, and $\mathbf{L}$ is the orbital angular momentum operator given by Eq. (A.3a). The ψ scalar functions are defined as

$$\psi^{\mathrm{L}} = \frac{1}{\nabla^2}\nabla \cdot \mathbf{E} , \tag{D.2a}$$

$$\psi^{\mathrm{M}} = \frac{1}{\mathrm{L}^2}\mathbf{L} \cdot \mathbf{E} , \tag{D.2b}$$

$$\psi^{\mathrm{E}} = \frac{ik}{\mathrm{L}^2\nabla^2}(\mathbf{L} \times \nabla) \cdot \mathbf{E} . \tag{D.2c}$$

The electric field satisfies the Helmholtz vector wave equation (2.24) at all points in space. As a consequence, the scalar functions satisfy the scalar Helmholtz equation

$$(\nabla^2 + k^2)\psi = 0 . \tag{D.3}$$

P. E. Stamatopoulou, *Strong Light-Matter Interactions in Extreme Plasmonic and Mie-Resonant Systems*, Springer Theses, https://doi.org/10.1007/978-3-031-67907-0

The longitudinal scalar function ψ^{L} describes instant propagation of the fields and contributes only in the cancellation of the pole of the TE wave close to the source region [7, 8], and we can therefore set $\psi^{\mathrm{L}} = 0$. The spherical symmetry of the system suggests the use of spherical coordinates. As already discussed in Appendices A, B and C, separation of variables to $\psi(r, \theta, \phi) = R(r)Y(\theta, \phi)$ allows us to write the overall solution as a linear combination of the spherical Bessel functions f_ℓ and spherical harmonics Y_ℓ^m, multiplied by appropriate expansion coefficients $c_{\ell m}$

$$\psi(r, \theta, \phi) = \sum_{\ell=1}^{\infty} \sum_{m=-\ell}^{\ell} c_{\ell m} f_\ell(kr) Y_\ell^m(\hat{\mathbf{r}}) \,. \tag{D.4}$$

By inserting Eq. (D.4) and $\psi^{\mathrm{L}} = 0$ back into Eq. (D.1) we obtain

$$\mathbf{E}(\mathbf{r}) = \mathbf{L} \sum_{\ell=1}^{\infty} \sum_{m=-\ell}^{\ell} b_{\ell m} f_\ell(kr) Y_\ell^m(\hat{\mathbf{r}}) + \frac{i}{k} \nabla \times \mathbf{L} \sum_{\ell=1}^{\infty} \sum_{m=-\ell}^{\ell} a_{\ell m} f_\ell(kr) Y_\ell^m(\hat{\mathbf{r}}) \,. \tag{D.5}$$

By introducing the vector spherical harmonics, defined as in Eq. (C.22), the multipolar expansion of the electric field takes the form

$$\mathbf{E}(\mathbf{r}) = \sum_{\ell m} \left\{ b_{\ell m} f_\ell(kr) \mathbf{X}_{\ell m}(\hat{\mathbf{r}}) + \frac{i}{k} a_{\ell m} \nabla \times f_\ell(kr) \mathbf{X}_{\ell m}(\hat{\mathbf{r}}) \right\} , \tag{D.6}$$

where we suppresed the notation of the double sum for brevity. Then the magnetic field can be obtained from Faraday's law (2.17c), yielding

$$\mathbf{H}(\mathbf{r}) = \frac{1}{Z} \sum_{\ell m} \left\{ a_{\ell m} f_\ell(kr) \mathbf{X}_{\ell m}(\hat{\mathbf{r}}) - \frac{i}{k} b_{\ell m} \nabla \times f_\ell(kr) \mathbf{X}_{\ell m}(\hat{\mathbf{r}}) \right\} , \tag{D.7}$$

where $Z = \sqrt{\mu\mu_0/(\varepsilon\varepsilon_0)}$ is the wave impedance.

In Eqs. (D.6) and (D.7) the function f_ℓ stands for any spherical Bessel-type function. In the general case the solution must encompass any combination of incoming and outgoing spherical waves

$$f_\ell = C_{\ell m}^{(+)} h_\ell^+ + C_{\ell m}^{(-)} h_\ell^- \,, \tag{D.8}$$

represented by the spherical Hankel function of the first, h_ℓ^+, and the second kind, h_ℓ^-, respectively. Given the restriction of finiteness of the field at $r \to 0$, inside the sphere this combination must be that of a standing wave $f_\ell = h_\ell^+ + i h_\ell^- = j_\ell$, so that the divergence $h_\ell^\pm(r \to 0) \to \infty$ cancels out. Therefore, the field inside the NP is

$$\mathbf{E}_{\mathrm{in}}(\mathbf{r}) = \sum_{\ell m} \left\{ b^{\mathrm{I}}_{\ell m} j_\ell(k_1 r)\mathbf{X}_{\ell m}(\hat{\mathbf{r}}) + \frac{i}{k_1} a^{\mathrm{I}}_{\ell m} \nabla \times j_\ell(k_1 r)\mathbf{X}_{\ell m}(\hat{\mathbf{r}}) \right\}, \tag{D.9a}$$

$$\mathbf{H}_{\mathrm{in}}(\mathbf{r}) = \frac{1}{Z_1} \sum_{\ell m} \left\{ a^{\mathrm{I}}_{\ell m} j_\ell(k_1 r)\mathbf{X}_{\ell m}(\hat{\mathbf{r}}) - \frac{i}{k_1} b^{\mathrm{I}}_{\ell m} \nabla \times j_\ell(k_1 r)\mathbf{X}_{\ell m}(\hat{\mathbf{r}}) \right\}, \tag{D.9b}$$

where index 1 refers to the optical parameters (ε_1, μ_1) of the medium in region I. The electric field in the host medium (ε_2, μ_2) consist of the incident plane wave $\mathbf{E}_{\mathrm{inc}}$ and a scattered wave $\mathbf{E}_{\mathrm{scat}}$

$$\mathbf{E}_{\mathrm{out}}(\mathbf{r}) = \mathbf{E}_{\mathrm{inc}}(\mathbf{r}) + \mathbf{E}_{\mathrm{scat}}(\mathbf{r}) . \tag{D.10}$$

The latter propagates outwards and is thus represented by a spherical Hankel function of the first kind

$$\mathbf{E}_{\mathrm{scat}}(\mathbf{r}) = \sum_{\ell m} \left\{ b^{\mathrm{II}}_{\ell m} h^+_\ell(k_2 r)\mathbf{X}_{\ell m}(\hat{\mathbf{r}}) + \frac{i}{k_2} a^{\mathrm{II}}_{\ell m} \nabla \times h^+_\ell(k_2 r)\mathbf{X}_{\ell m}(\hat{\mathbf{r}}) \right\}, \tag{D.11a}$$

$$\mathbf{H}_{\mathrm{scat}}(\mathbf{r}) = \frac{1}{Z_2} \sum_{\ell m} \left\{ a^{\mathrm{II}}_{\ell m} h^+_\ell(k_2 r)\mathbf{X}_{\ell m}(\hat{\mathbf{r}}) - \frac{i}{k_2} b^{\mathrm{II}}_{\ell m} \nabla \times h^+_\ell(k_2 r)\mathbf{X}_{\ell m}(\hat{\mathbf{r}}) \right\}. \tag{D.11b}$$

A plane EM wave of the form of Eq. (2.30), can be expanded in spherical standing waves as [4]

$$\mathrm{e}^{i\mathbf{k}\cdot\mathbf{r}} = 4\pi \sum_{\ell=0}^{\infty} \sum_{m=-\ell}^{\ell} i^\ell j_\ell(kr) Y^m_\ell(\hat{\mathbf{r}}) Y^{m*}_\ell(\hat{\mathbf{k}}) , \tag{D.12}$$

which allows us to write the incident field as

$$\mathbf{E}_{\mathrm{inc}}(\mathbf{r}) = \sum_{\ell m} \left\{ b^{0}_{\ell m} j_\ell(k_2 r)\mathbf{X}_{\ell m}(\hat{\mathbf{r}}) + \frac{i}{k_2} a^{0}_{\ell m} \nabla \times j_\ell(k_2 r)\mathbf{X}_{\ell m}(\hat{\mathbf{r}}) \right\}, \tag{D.13a}$$

$$\mathbf{H}_{\mathrm{inc}}(\mathbf{r}) = \frac{1}{Z_2} \sum_{\ell m} \left\{ a^{0}_{\ell m} j_\ell(k_2 r)\mathbf{X}_{\ell m}(\hat{\mathbf{r}}) - \frac{i}{k_2} b^{0}_{\ell m} \nabla \times j_\ell(k_2 r)\mathbf{X}_{\ell m}(\hat{\mathbf{r}}) \right\}. \tag{D.13b}$$

Notice that the 0-order term is eliminated, since $\mathbf{X}_{00}(\hat{\mathbf{r}}) = 0$, reflecting the absence of monopoles in CED. To determine the coefficients $a^0_{\ell m}$ and $b^0_{\ell m}$ we insert Eqs. (D.13) into Eqs. (2.30), and using the orthonormality relation of the vector spherical harmonics (see Appendix C) we find

$$
\begin{aligned}
a^0_{\ell m} = \frac{4\pi i^\ell(-1)^{m+1}}{\sqrt{\ell(\ell+1)}} \Big\{ & \big[\alpha_\ell^m Y_\ell^{-(m+1)}(\hat{\mathbf{k}}) + \alpha_\ell^{-m} Y_\ell^{-(m-1)}(\hat{\mathbf{k}})\big](\hat{\mathbf{k}}\times\mathbf{E}_0)_x \\
& + i\big[\alpha_\ell^m Y_\ell^{-(m+1)}(\hat{\mathbf{k}}) - \alpha_\ell^{-m} Y_\ell^{-(m-1)}(\hat{\mathbf{k}})\big](\hat{\mathbf{k}}\times\mathbf{E}_0)_y \\
& - m Y_\ell^{-m}(\hat{\mathbf{k}})(\hat{\mathbf{k}}\times\mathbf{E}_0)_z \Big\} ,
\end{aligned} \tag{D.14a}
$$

$$
\begin{aligned}
b^0_{\ell m} = \frac{4\pi i^\ell(-1)^{m+1}}{\sqrt{\ell(\ell+1)}} \Big\{ & \big[\alpha_\ell^m Y_\ell^{-(m+1)}(\hat{\mathbf{k}}) + \alpha_\ell^{-m} Y_\ell^{-(m-1)}(\hat{\mathbf{k}})\big]E_{0x} \\
& + i\big[\alpha_\ell^m Y_\ell^{-(m+1)}(\hat{\mathbf{k}}) - \alpha_\ell^{-m} Y_\ell^{-(m-1)}(\hat{\mathbf{k}})\big]E_{0y} \\
& - m Y_\ell^{-m}(\hat{\mathbf{k}})E_{0z} \Big\} ,
\end{aligned} \tag{D.14b}
$$

with $\alpha_\ell^m = \sqrt{(\ell-m)(\ell+m+1)}/2$ as in Eq. (C.5).

Mie Coefficients To determine $a^{\mathrm{I/II}}_{\ell m}$ and $b^{\mathrm{I/II}}_{\ell m}$, one has to impose the boundary conditions at the interface separating the NP from the surrounding medium. According to Eqs. (2.15), the tangential components of the electric and magnetic field are continuous across the boundary, giving

$$
\mathbf{X}^*_{\ell m}(\hat{\mathbf{r}})\cdot\mathbf{E}_{\mathrm{in}}(\mathbf{r})\Big|_{r=R} = \mathbf{X}^*_{\ell m}(\hat{\mathbf{r}})\cdot\mathbf{E}_{\mathrm{out}}(\mathbf{r})\Big|_{r=R} , \tag{D.15a}
$$

$$
\mathbf{X}^*_{\ell m}(\hat{\mathbf{r}})\cdot\mathbf{H}_{\mathrm{in}}(\mathbf{r})\Big|_{r=R} = \mathbf{X}^*_{\ell m}(\hat{\mathbf{r}})\cdot\mathbf{H}_{\mathrm{out}}(\mathbf{r})\Big|_{r=R} , \tag{D.15b}
$$

$$
\big[\hat{\mathbf{r}}\times\mathbf{X}^*_{\ell m}(\hat{\mathbf{r}})\big]\cdot\mathbf{E}_{\mathrm{in}}(\mathbf{r})\Big|_{r=R} = \big[\hat{\mathbf{r}}\times\mathbf{X}^*_{\ell m}(\hat{\mathbf{r}})\big]\cdot\mathbf{E}_{\mathrm{out}}(\mathbf{r})\Big|_{r=R} , \tag{D.15c}
$$

$$
\big[\hat{\mathbf{r}}\times\mathbf{X}^*_{\ell m}(\hat{\mathbf{r}})\big]\cdot\mathbf{H}_{\mathrm{in}}(\mathbf{r})\Big|_{r=R} = \big[\hat{\mathbf{r}}\times\mathbf{X}^*_{\ell m}(\hat{\mathbf{r}})\big]\cdot\mathbf{H}_{\mathrm{out}}(\mathbf{r})\Big|_{r=R} . \tag{D.15d}
$$

Using the orthogonality relations (C.31), as well as Eqs. (C.40), we find

$$
b^{\mathrm{I}}_{\ell m} j_\ell(k_1 R) = b^0_{\ell m} j_\ell(k_2 R) + b^{\mathrm{II}}_{\ell m} h^+_\ell(k_2 R) , \tag{D.16a}
$$

$$
\frac{1}{Z_1} a^{\mathrm{I}}_{\ell m} j_\ell(k_1 R) = \frac{1}{Z_2} a^0_{\ell m} j_\ell(k_2 R) + \frac{1}{Z_2} a^{\mathrm{II}}_{\ell m} h^+_\ell(k_2 R) , \tag{D.16b}
$$

$$
\frac{1}{k_1} a^{\mathrm{I}}_{\ell m} \Psi'_\ell(k_1 R) = \frac{1}{k_2} a^0_{\ell m} \Psi'_\ell(k_2 R) + \frac{1}{k_2} a^{\mathrm{II}}_{\ell m} \xi'_\ell(k_2 R) , \tag{D.16c}
$$

$$
\frac{1}{Z_1 k_1} b^{\mathrm{I}}_{\ell m} \Psi'_\ell(k_1 R) = \frac{1}{Z_2 k_2} b^0_{\ell m} \Psi'_\ell(k_2 R) + \frac{1}{Z_2 k_2} b^{\mathrm{II}}_{\ell m} \xi'_\ell(k_2 R) , \tag{D.16d}
$$

where we adopted the notation of the Riccati-Bessel functions $\Psi_\ell(x) = x j_\ell(x)$ and $\xi_\ell(x) = x h^+_\ell(x)$. The prime denotes the functions' derivatives with respect to their argument, namely $f'_\ell(k_i R) = d_x f_\ell(x)|_{x=k_i R}$. From the set of equations above we obtain the Mie coefficients linking the incident TE and TM waves to the respective scattered ones, as

$$T_{E\ell} = \frac{a^{II}_{\ell m}}{a^{0}_{\ell m}} = \frac{\varepsilon_1 j_\ell(k_1R)\Psi'_\ell(k_2R) - \varepsilon_2\Psi'_\ell(k_1R)j_\ell(k_2R)}{\varepsilon_2 h^+_\ell(k_2R)\Psi'_\ell(k_1R) - \varepsilon_1\xi'_\ell(k_2R)j_\ell(k_1R)}, \tag{D.17a}$$

$$T_{M\ell} = \frac{b^{II}_{\ell m}}{b^{0}_{\ell m}} = \frac{\mu_1 j_\ell(k_1R)\Psi'_\ell(k_2R) - \mu_2\Psi'_\ell(k_1R)j_\ell(k_2R)}{\mu_2 h^+_\ell(k_2R)\Psi'_\ell(k_1R) - \mu_1\xi'_\ell(k_2R)j_\ell(k_1R)}, \tag{D.17b}$$

and to the internal TE and TM waves, as

$$T^{21}_{E\ell} = \frac{a^{I}_{\ell m}}{a^{0}_{\ell m}} = \frac{\varepsilon_2 k_1}{k_2}\frac{h^+_\ell(k_2R)\Psi'_\ell(k_2R) - \xi'_\ell(k_2R)j_\ell(k_2R)}{\varepsilon_2 h^+_\ell(k_2R)\Psi'_\ell(k_1R) - \varepsilon_1\xi'_\ell(k_2R)j_\ell(k_1R)}, \tag{D.18a}$$

$$T^{21}_{M\ell} = \frac{b^{I}_{\ell m}}{b^{0}_{\ell m}} = \mu_1\frac{h^+_\ell(k_2R)\Psi'_\ell(k_2R) - \xi'_\ell(k_2R)j_\ell(k_2R)}{\mu_2 h^+_\ell(k_2R)\Psi'_\ell(k_1R) - \mu_1\xi'_\ell(k_2R)j_\ell(k_1R)}. \tag{D.18b}$$

Using the Wronskian determinant of Eq. (B.10a), the latter relations can be written as

$$T^{21}_{E\ell} = -\frac{i}{k_2R}\sqrt{\frac{\varepsilon_1\varepsilon_2\mu_1}{\mu_2}}\frac{1}{\varepsilon_2 h^+_\ell(k_2R)\Psi'_\ell(k_1R) - \varepsilon_1\xi'_\ell(k_2R)j_\ell(k_1R)}, \tag{D.19a}$$

$$T^{21}_{M\ell} = -\frac{i\mu_1}{k_2R}\frac{1}{\mu_2 h^+_\ell(k_2R)\Psi'_\ell(k_1R) - \mu_1\xi'_\ell(k_2R)j_\ell(k_1R)}. \tag{D.19b}$$

Here the superscript 21 indicates that the coefficients relate the incident fields in medium 2 to the ones induced in medium 1.

Optical Cross Sections To quantify the scattering and absorption properties of a NP we typically consider the scattering and absorption powers and cross sections. The scattering power corresponds to the energy flow,[2] of the scattered field out of the boundary outlined by the NP surface S[3] whereas the absorption power expresses the energy inflow of the total fields into the boundary. For time-harmonic fields both powers can be calculated by the time-averaged Poynting vector $\langle \mathbf{S}\rangle = \mathrm{Re}\{\mathbf{E}\times\mathbf{H}^*\}/2$ integrated over the boundary

$$P_{\mathrm{scat}} = \frac{R^2}{2}\mathrm{Re}\int_S d\Omega\big[\mathbf{E}_{\mathrm{scat}}(\mathbf{r})\times\mathbf{H}^*_{\mathrm{scat}}(\mathbf{r})\big]\cdot\hat{\mathbf{r}}, \tag{D.20a}$$

$$P_{\mathrm{abs}} = -\frac{R^2}{2}\mathrm{Re}\int_S d\Omega\big[\mathbf{E}_{\mathrm{out}}(\mathbf{r})\times\mathbf{H}^*_{\mathrm{out}}(\mathbf{r})\big]\cdot\hat{\mathbf{r}}, \tag{D.20b}$$

where the negative sign in the definition of P_{abs} indicates inward direction of energy flux, opposite to the normal (to the boundary) vector $\hat{\mathbf{r}}$. In practice, we may evaluate the Poynting vector in the far field ($kr\to\infty$), rather than the surface of the sphere, since there is no creation or destruction of charges in between. This allows to use the

[2] The energy flow of an EM field per unit area per unit time is described by the Poynting vector $\mathbf{S} = \mathbf{E}\times\mathbf{H}$.

[3] Not to be confused with the Poynting vector $\mathbf{S}$.

asymptotic behavior of the spherical Bessel functions of Eqs. (B.8) in the expressions of the fields. By substituting the scattered field formulas (D.11) in Eq. (D.20a) and the total fields of Eqs. (D.10), (D.11), (D.13) in Eq. (D.20b), and after a considerable amount of algebra, we find

$$P_{\text{scat}} = \frac{1}{2k_2^2 Z_2} \sum_{\ell m} \left\{ \left| a_{\ell m}^{\text{II}} \right|^2 + \left| b_{\ell m}^{\text{II}} \right|^2 \right\}, \tag{D.21a}$$

$$P_{\text{abs}} = -\frac{1}{2k_2^2 Z_2} \sum_{\ell m} \left(\left| a_{\ell m}^{\text{II}} \right|^2 + \left| b_{\ell m}^{\text{II}} \right|^2 + \text{Re}\left\{ T_{\text{E}\ell} \left| a_{\ell m}^{0} \right|^2 + T_{\text{M}\ell} \left| b_{\ell m}^{0} \right|^2 \right\} \right). \tag{D.21b}$$

For the calculation above, one may find useful the integrals given in Eqs. (C.40). Assuming, without loss of generality, propagation along the z direction ($\theta = 0$), and with use of Eq. (C.15), the summation over m yields

$$\sum_{m=-\ell}^{\ell} |a_{\ell m}^{0}|^2 = \sum_{m=-\ell}^{\ell} |b_{\ell m}^{0}|^2 = 2\pi(2\ell+1)|\mathbf{E}_0|^2 , \tag{D.22}$$

so that

$$P_{\text{scat}} = \frac{\pi |\mathbf{E}_0|^2}{k_2^2 Z_2} \sum_{\ell} (2\ell+1) \left\{ \left| T_{\text{E}\ell} \right|^2 + \left| T_{\text{M}\ell} \right|^2 \right\}, \tag{D.23a}$$

$$P_{\text{abs}} = -\frac{\pi |\mathbf{E}_0|^2}{k_2^2 Z_2} \sum_{\ell} (2\ell+1) \left(\left| T_{\text{E}\ell} \right|^2 + \left| T_{\text{M}\ell} \right|^2 + \text{Re}\left\{ T_{\text{E}\ell} + T_{\text{M}\ell} \right\} \right). \tag{D.23b}$$

The sum of the scattering and absorption power gives, naturally, the extinction power

$$P_{\text{ext}} = -\frac{\pi |\mathbf{E}_0|^2}{k_2^2 Z_2} \sum_{\ell} (2\ell+1) \text{Re}\left\{ T_{\text{E}\ell} + T_{\text{M}\ell} \right\}. \tag{D.24}$$

The scattering, absorption and extinction cross sections $\sigma_{\text{scat/abs/ext}}$ are defined as the ratio of the corresponding power over the power of the incident plane wave $P_0 = \mathbf{S}_0 \cdot \hat{\mathbf{k}} = |\mathbf{E}_0|^2/(2Z_2)$, yielding

$$\sigma_{\text{scat}} = \frac{2\pi}{k_2^2} \sum_{\ell} (2\ell+1) \left\{ \left| T_{\text{E}\ell} \right|^2 + \left| T_{\text{M}\ell} \right|^2 \right\}, \tag{D.25a}$$

$$\sigma_{\text{abs}} = -\frac{2\pi}{k_2^2} \sum_{\ell} (2\ell+1) \left(\left| T_{\text{E}\ell} \right|^2 + \left| T_{\text{M}\ell} \right|^2 + \text{Re}\left\{ T_{\text{E}\ell} + T_{\text{M}\ell} \right\} \right), \tag{D.25b}$$

$$\sigma_{\text{ext}} = -\frac{2\pi}{k_2^2} \sum_{\ell} (2\ell+1) \text{Re}\left\{ T_{\text{E}\ell} + T_{\text{M}\ell} \right\}. \tag{D.25c}$$

Finally, the optical cross sections appear typically normalized to the geometrical cross section πR^2

$$\sigma_{\mathrm{scat}} = \frac{2}{(k_2 R)^2} \sum_{\ell} (2\ell + 1) \left(\left|T_{\mathrm{E}\ell}\right|^2 + \left|T_{\mathrm{M}\ell}\right|^2 \right) , \quad \text{(D.26a)}$$

$$\sigma_{\mathrm{abs}} = -\frac{2}{(k_2 R)^2} \sum_{\ell} (2\ell + 1) \left(\left|T_{\mathrm{E}\ell}\right|^2 + \left|T_{\mathrm{M}\ell}\right|^2 + \mathrm{Re}\left\{T_{\mathrm{E}\ell} + T_{\mathrm{M}\ell}\right\} \right) , \quad \text{(D.26b)}$$

$$\sigma_{\mathrm{ext}} = -\frac{2}{(k_2 R)^2} \sum_{\ell} (2\ell + 1) \mathrm{Re}\left\{T_{\mathrm{E}\ell} + T_{\mathrm{M}\ell}\right\} . \quad \text{(D.26c)}$$

These dimensionless cross sections go often under the name *optical efficiencies* in literature [9]. Throughout the present thesis we use the term 'cross section' to refer to the normalized quantities of Eqs. (D.26).

D.2 Core–Shell Geometry

Following the basic steps of Mie theory, one can readily derive analytic formulas for the optical cross sections of NPs consisting of multiple concentric shells under plane-wave illumination. Here we focus on core–shell NPs like the one depicted in Fig. 3.1 [9], which consist of a simple isotropic spherical NP of radius R_1 (region I) and a concentric shell of outer radius R_2 (region II), suspended in a non-absorbing host medium (region III). In the formalism that follows, labels 1, 2, 3 refer to the parameters pertaining to the core, shell, and host medium accordingly.

Electromagnetic Fields In the same spirit as in Eqs. (D.9), (D.11) and (D.13), the EM fields in each of the three regions take the form

$$\mathbf{E}_{\mathrm{I}}(\mathbf{r}) = \sum_{\ell m} \left\{ b^{\mathrm{I}}_{\ell m} j_\ell(k_1 r) \mathbf{X}_{\ell m}(\hat{\mathbf{r}}) + \frac{i}{k_1} a^{\mathrm{I}}_{\ell m} \nabla \times j_\ell(k_1 r) \mathbf{X}_{\ell m}(\hat{\mathbf{r}}) \right\} , \quad \text{(D.27a)}$$

$$\mathbf{H}_{\mathrm{I}}(\mathbf{r}) = \frac{1}{Z_1} \sum_{\ell m} \left\{ a^{\mathrm{I}}_{\ell m} j_\ell(k_1 r) \mathbf{X}_{\ell m}(\hat{\mathbf{r}}) - \frac{i}{k_1} b^{\mathrm{I}}_{\ell m} \nabla \times j_\ell(k_1 r) \mathbf{X}_{\ell m}(\hat{\mathbf{r}}) \right\} , \quad \text{(D.27b)}$$

$$\begin{aligned} \mathbf{E}_{\mathrm{II}}(\mathbf{r}) = \sum_{\ell m} \Big\{ & b^{\mathrm{II}}_{\ell m} h^+_\ell(k_2 r) \mathbf{X}_{\ell m}(\hat{\mathbf{r}}) + \frac{i}{k_2} a^{\mathrm{II}}_{\ell m} \nabla \times h^+_\ell(k_2 r) \mathbf{X}_{\ell m}(\hat{\mathbf{r}}) \\ & + b^{\mathrm{II},0}_{\ell m} j_\ell(k_2 r) \mathbf{X}_{\ell m}(\hat{\mathbf{r}}) + \frac{i}{k_2} a^{\mathrm{II},0}_{\ell m} \nabla \times j_\ell(k_2 r) \mathbf{X}_{\ell m}(\hat{\mathbf{r}}) \Big\} , \end{aligned} \quad \text{(D.27c)}$$

$$\mathbf{H}_{\mathrm{II}}(\mathbf{r}) = \frac{1}{Z_2} \sum_{\ell m} \Big\{ a^{\mathrm{II}}_{\ell m} h^+_\ell(k_2 r) \mathbf{X}_{\ell m}(\hat{\mathbf{r}}) - \frac{i}{k_2} b^{\mathrm{II}}_{\ell m} \nabla \times h^+_\ell(k_2 r) \mathbf{X}_{\ell m}(\hat{\mathbf{r}})$$

$$+ a_{\ell m}^{\mathrm{II},0} j_\ell(k_2 r)\mathbf{X}_{\ell m}(\hat{\mathbf{r}}) - \frac{i}{k_2} b_{\ell m}^{\mathrm{II},0} \nabla \times j_\ell(k_2 r)\mathbf{X}_{\ell m}(\hat{\mathbf{r}})\Big\} , \tag{D.27d}$$

$$\mathbf{E}_{\mathrm{III}}(\mathbf{r}) = \sum_{\ell m} \Big\{ b_{\ell m}^{\mathrm{III}} h_\ell^+(k_3 r)\mathbf{X}_{\ell m}(\hat{\mathbf{r}}) + \frac{i}{k_3} a_{\ell m}^{\mathrm{III}} \nabla \times h_\ell^+(k_3 r)\mathbf{X}_{\ell m}(\hat{\mathbf{r}})$$

$$+ b_{\ell m}^{0} j_\ell(k_3 r)\mathbf{X}_{\ell m}(\hat{\mathbf{r}}) + \frac{i}{k_3} a_{\ell m}^{0} \nabla \times j_\ell(k_3 r)\mathbf{X}_{\ell m}(\hat{\mathbf{r}})\Big\} , \tag{D.27e}$$

$$\mathbf{H}_{\mathrm{III}}(\mathbf{r}) = \frac{1}{Z_3} \sum_{\ell m} \Big\{ a_{\ell m}^{\mathrm{III}} h_\ell^+(k_3 r)\mathbf{X}_{\ell m}(\hat{\mathbf{r}}) - \frac{i}{k_3} b_{\ell m}^{\mathrm{III}} \nabla \times h_\ell^+(k_3 r)\mathbf{X}_{\ell m}(\hat{\mathbf{r}})$$

$$+ a_{\ell m}^{0} j_\ell(k_3 r)\mathbf{X}_{\ell m}(\hat{\mathbf{r}}) - \frac{i}{k_3} b_{\ell m}^{0} \nabla \times j_\ell(k_3 r)\mathbf{X}_{\ell m}(\hat{\mathbf{r}})\Big\} , \tag{D.27f}$$

with $a/b_{\ell m}^0$ given by Eqs. (D.14), while the remaining coefficients are determined via the boundary conditions.

Mie Coefficients By invoking the continuity of the tangential EM fields at the boundary between the core and the shell (at $r = R_1$) we obtain a system of equations identical to (D.16), whereas at the outermost boundary (at $r = R_2$), we find

$$b_{\ell m}^{\mathrm{II},0} j_\ell(k_2 R_2) + b_{\ell m}^{\mathrm{II}} h_\ell^+(k_2 R_2) = b_{\ell m}^{0} j_\ell(k_3 R_2) + b_{\ell m}^{\mathrm{III}} h_\ell^+(k_3 R_2) , \tag{D.28a}$$

$$\frac{1}{Z_2} a_{\ell m}^{\mathrm{II},0} j_\ell(k_2 R_2) + \frac{1}{Z_2} a_{\ell m}^{\mathrm{II}} h_\ell^+(k_2 R_2) = \frac{1}{Z_3} a_{\ell m}^{0} j_\ell(k_3 R_2) + \frac{1}{Z_3} a_{\ell m}^{\mathrm{III}} h_\ell^+(k_3 R_2) , \tag{D.28b}$$

$$\frac{1}{k_2} a_{\ell m}^{\mathrm{II},0} \Psi_\ell'(k_2 R_2) + \frac{1}{k_2} a_{\ell m}^{\mathrm{II}} \xi_\ell'(k_2 R_2) = \frac{1}{k_3} a_{\ell m}^{0} \Psi_\ell'(k_3 R_2) + \frac{1}{k_3} a_{\ell m}^{\mathrm{III}} \xi_\ell'(k_3 R_2) , \tag{D.28c}$$

$$\frac{1}{Z_2 k_2} b_{\ell m}^{\mathrm{II},0} \Psi_\ell'(k_2 R_2) + \frac{1}{Z_2 k_2} b_{\ell m}^{\mathrm{II}} \xi_\ell'(k_2 R_2) = \frac{1}{Z_3 k_3} b_{\ell m}^{0} \Psi_\ell'(k_3 R_2) + \frac{1}{Z_3 k_3} b_{\ell m}^{\mathrm{III}} \xi_\ell'(k_3 R_2) . \tag{D.28d}$$

The two sets of equations lead to the following connection relations between the different expansion coefficients:

$$a/b_{\ell m}^{\mathrm{III}} = T_{\mathrm{E/M}\ell}\, a/b_{\ell m}^0 , \tag{D.29a}$$

$$a/b_{\ell m}^{\mathrm{II},0} = T_{\mathrm{E/M}\ell}^{32}\, a/b_{\ell m}^0 , \tag{D.29b}$$

$$a/b_{\ell m}^{\mathrm{II}} = T_{\mathrm{E/M}\ell}^{22}\, a/b_{\ell m}^{\mathrm{II},0} = T_{\mathrm{E/M}\ell}^{22} T_{\mathrm{E/M}\ell}^{32}\, a/b_{\ell m}^0 , \tag{D.29c}$$

$$a/b_{\ell m}^{\mathrm{I}} = T_{\mathrm{E/M}\ell}^{21}\, a/b_{\ell m}^{\mathrm{II},0} = T_{\mathrm{E/M}\ell}^{21} T_{\mathrm{E/M}\ell}^{32}\, a/b_{\ell m}^0 . \tag{D.29d}$$

In Eqs. (D.29), $T_{\mathrm{E/M}\ell}^{22}$ and $T_{\mathrm{E/M}\ell}^{21}$ correspond to the Mie coefficients for the scattered field and internal field, respectively, of the uncoated NP, as given by Eqs. (D.17) and (D.19). The Mie coefficients linking the incident field to the one scattered at the outer surface of the core–shell NP read

$$T_{\mathrm{E}\ell} = \frac{\left[\varepsilon_2 h_\ell^+(x_2)\Psi_\ell'(x_3) - \varepsilon_3 \xi_\ell'(x_2) j_\ell(x_3)\right] T_{\mathrm{E}\ell}^{22} + \varepsilon_2 j_\ell(x_2)\Psi_\ell'(x_3) - \varepsilon_3 \Psi_\ell'(x_2) j_\ell(x_3)}{\left[\varepsilon_3 h_\ell^+(x_3)\xi_\ell'(x_2) - \varepsilon_2 \xi_\ell'(x_3) h_\ell^+(x_2)\right] T_{\mathrm{E}\ell}^{22} + \varepsilon_3 h_\ell^+(x_3)\Psi_\ell'(x_2) - \varepsilon_2 \xi_\ell'(x_3) j_\ell(x_2)} , \tag{D.30a}$$

$$T_{\mathrm{M}\ell} = \frac{\left[\mu_2 h_\ell^+(x_2)\Psi_\ell'(x_3) - \mu_3\xi_\ell'(x_2)j_\ell(x_3)\right]T_{\mathrm{M}\ell}^{22} + \mu_2 j_\ell(x_2)\Psi_\ell'(x_3) - \mu_3\Psi_\ell'(x_2)j_\ell(x_3)}{\left[\mu_3 h_\ell^+(x_3)\xi_\ell'(x_2) - \mu_2\xi_\ell'(x_3)h_\ell^+(x_2)\right]T_{\mathrm{M}\ell}^{22} + \mu_3 h_\ell^+(x_3)\Psi_\ell'(x_2) - \mu_2\xi_\ell'(x_3)j_\ell(x_2)}. \tag{D.30b}$$

where we have adopted the notation $x_i = k_i R_2$. The fields in the shell are linked to the incident via the following relations

$$T_{\mathrm{E}\ell}^{32} = \frac{-i\sqrt{\varepsilon_2\varepsilon_3\mu_2/\mu_3}/x_3}{\left[\varepsilon_3 h_\ell^+(x_3)\xi_\ell'(x_2) - \varepsilon_2\xi_\ell'(x_3)h_\ell^+(x_2)\right]T_{\mathrm{E}\ell}^{22} + \varepsilon_3 h_\ell^+(x_3)\Psi_\ell'(x_2) - \varepsilon_2\xi_\ell'(x_3)j_\ell(x_2)}, \tag{D.31a}$$

$$T_{\mathrm{M}\ell}^{32} = \frac{-i\mu_2/x_3}{\left[\mu_3 h_\ell^+(x_3)\xi_\ell'(x_2) - \mu_2\xi_\ell'(x_3)h_\ell^+(x_2)\right]T_{\mathrm{M}\ell}^{22} + \mu_3 h_\ell^+(x_3)\Psi_\ell'(x_2) - \mu_2\xi_\ell'(x_3)j_\ell(x_2)}, \tag{D.31b}$$

The optical cross sections can be now calculated by inserting the Mie coefficients of Eqs. (D.30) into Eqs. (D.26).

D.3 Chiral Medium

Let us now consider a chiral NP, described by optical parameters (ε_1, μ_1, κ), with κ being the chirality parameter, and excited by a circularly polarized plane wave. The fields inside the NP now obey the constitutive relations of Eqs. (2.41). From the Ampère-Maxwell and the Faraday law we can construct the following equation

$$\nabla \times \begin{pmatrix}\mathbf{E}\\ \mathbf{H}\end{pmatrix} = \mathbf{K}\begin{pmatrix}\mathbf{E}\\ \mathbf{H}\end{pmatrix}, \tag{D.32}$$

with

$$\mathbf{K} = \begin{pmatrix}\kappa\omega/c & i\omega\mu_1\mu_0\\ -i\omega\varepsilon_1\varepsilon_0 & \kappa\omega/c\end{pmatrix}. \tag{D.33}$$

As mentioned in Sect. 2.2, only circularly polarized waves can propagate inside the medium without suffering OR. It is then natural to express the field inside the NP as a combination of LCP (+) and RCP (−) waves, say $\mathbf{Q}_\mathrm{L}$ and $\mathbf{Q}_\mathrm{R}$. A linear transformation $\mathbf{A}$ of the EM field

$$\begin{pmatrix}\mathbf{E}\\ \mathbf{H}\end{pmatrix} = \mathbf{A}\begin{pmatrix}\mathbf{Q}_\mathrm{L}\\ \mathbf{Q}_\mathrm{R}\end{pmatrix} \tag{D.34}$$

diagonalizes $\mathbf{K}$ such that $\Lambda = \mathbf{A}^{-1}\mathbf{K}\mathbf{A}$. This holds for

$$\Lambda = \begin{pmatrix}k_\mathrm{L} & 0\\ 0 & -k_\mathrm{R}\end{pmatrix}, \quad \text{and} \quad \mathbf{A} = \begin{pmatrix}1 & 1\\ -i/Z_1 & +i/Z_1\end{pmatrix}, \tag{D.35}$$

where $k_{L/R} = k_1 \pm \kappa\omega/c$ are the wavenumbers related to the LCP/RCP waves. The corresponding refractive indices are $n_{L/R} = n_1 \pm \kappa$, with $(n_L + n_R)/2 = n_1 = \sqrt{\varepsilon_1\mu_1}$ expressing the average (background) refractive index. Indeed, one can see from

$$i\hat{\mathbf{k}}k_{L/R} \times \mathbf{Q}_{L/R} = \pm k_{L/R}\mathbf{Q}_{L/R} \quad \text{and} \quad \hat{\mathbf{k}} \cdot \mathbf{Q}_{L/R} = 0\,, \tag{D.36}$$

that they take the form $\mathbf{Q}_{L/R} = Q(\hat{\mathbf{e}}_1 \pm i\hat{\mathbf{e}}_2)$, thus describing LCP and RCP waves.

According to Eq. (D.34) the electric and magnetic field inside the chiral medium take the general form

$$\mathbf{E} = \mathbf{Q}_L + \mathbf{Q}_R\,, \tag{D.37a}$$

$$\mathbf{H} = -\frac{i}{Z_1}\left(\mathbf{Q}_L - \mathbf{Q}_R\right)\,. \tag{D.37b}$$

We can therefore express the fields inside the NP as

$$\mathbf{E}_{\text{in}}(\mathbf{r}) = \sum_{\ell m}\Bigg\{c^{L}_{\ell m}\left[j_\ell(k_L r)\mathbf{X}_{\ell m}(\hat{\mathbf{r}}) + \frac{1}{k_L}\nabla \times j_\ell(k_L r)\mathbf{X}_{\ell m}(\hat{\mathbf{r}})\right] + c^{R}_{\ell m}\left[j_\ell(k_R r)\mathbf{X}_{\ell m}(\hat{\mathbf{r}}) - \frac{1}{k_R}\nabla \times j_\ell(k_R r)\mathbf{X}_{\ell m}(\hat{\mathbf{r}})\right]\Bigg\}\,, \tag{D.38a}$$

$$\mathbf{H}_{\text{in}}(\mathbf{r}) = \frac{i}{Z_1}\sum_{\ell m}\Bigg\{-c^{L}_{\ell m}\left[j_\ell(k_L r)\mathbf{X}_{\ell m}(\hat{\mathbf{r}}) + \frac{1}{k_L}\nabla \times j_\ell(k_L r)\mathbf{X}_{\ell m}(\hat{\mathbf{r}})\right] + c^{R}_{\ell m}\left[j_\ell(k_R r)\mathbf{X}_{\ell m}(\hat{\mathbf{r}}) - \frac{i}{k_R}\nabla \times j_\ell(k_R r)\mathbf{X}_{\ell m}(\hat{\mathbf{r}})\right]\Bigg\}\,, \tag{D.38b}$$

whereas the fields in the surrounding medium are the same as in the isotropic medium case. Following the usual steps of Mie theory (requirement of continuity of the tangential fields at the boundary), we find that the amplitudes of the scattered fields $a^{II}_{\ell m}$ and $b^{II}_{\ell m}$ are connected to the incident amplitudes $a^{0}_{\ell m}$ and $b^{0}_{\ell m}$, via

$$\begin{pmatrix} a^{II}_{\ell m} \\ b^{II}_{\ell m} \end{pmatrix} = \begin{pmatrix} T_{EE\ell} & T_{EM\ell} \\ T_{ME\ell} & T_{MM\ell} \end{pmatrix}\begin{pmatrix} a^{0}_{\ell m} \\ b^{0}_{\ell m} \end{pmatrix}\,. \tag{D.39}$$

The elements of the T matrix are given by

$$T_{EE\ell} = \frac{A(L)V(R) + A(R)V(L)}{W(L)V(R) + W(R)V(L)} = T_{E\ell}\,, \tag{D.40a}$$

$$T_{MM\ell} = \frac{B(L)W(R) + B(R)W(L)}{W(L)V(R) + W(R)V(L)} = T_{M\ell}\,, \tag{D.40b}$$

$$T_{\mathrm{ME}\ell} = i\frac{A(L)W(R) - A(R)W(L)}{W(L)V(R) + W(R)V(L)} = T_{\mathrm{C}\ell}\,, \tag{D.40c}$$

$$T_{\mathrm{EM}\ell} = -i\frac{B(L)V(R) - B(R)V(L)}{W(L)V(R) + W(R)V(L)} = -T_{\mathrm{C}\ell}\,, \tag{D.40d}$$

where we have set for $\mathrm{P} = \mathrm{L}, \mathrm{R}$

$$A(\mathrm{P}) = \frac{1}{Z_1 k_2} j_\ell(k_\mathrm{P} R)\Psi'_\ell(k_2 R) - \frac{1}{Z_2 k_\mathrm{P}}\Psi'_\ell(k_\mathrm{P} R) j_\ell(k_2 R)\,, \tag{D.41a}$$

$$B(\mathrm{P}) = \frac{1}{Z_2 k_2} j_\ell(k_\mathrm{P} R)\Psi'_\ell(k_2 R) - \frac{1}{Z_1 k_\mathrm{P}}\Psi'_\ell(k_\mathrm{P} R) j_\ell(k_2 R)\,, \tag{D.41b}$$

$$V(\mathrm{P}) = \frac{1}{Z_1 k_\mathrm{P}} h_\ell^+(k_2 R)\Psi'_\ell(k_\mathrm{P} R) - \frac{1}{Z_2 k_2}\xi'_\ell(k_2 R) j_\ell(k_\mathrm{P} R)\,, \tag{D.41c}$$

$$W(\mathrm{P}) = \frac{1}{Z_2 k_\mathrm{P}} h_\ell^+(k_2 R)\Psi'_\ell(k_\mathrm{P} R) - \frac{1}{Z_1 k_2}\xi'_\ell(k_2 R) j_\ell(k_\mathrm{P} R)\,. \tag{D.41d}$$

Then the optical cross sections (normalized to the geometrical cross section) can be evaluated by Eqs. (D.21), i.e.

$$\sigma_{\mathrm{scat}} = \frac{1}{(k_2 R)^2 \pi |\mathbf{E}_0|^2} \sum_{\ell m} \left(|a_{\ell m}^{\mathrm{II}}|^2 + |b_{\ell m}^{\mathrm{II}}|^2 \right), \tag{D.42a}$$

$$\begin{aligned}\sigma_{\mathrm{ext}} = -\frac{1}{(k_2 R)\pi |\mathbf{E}_0|^2} \mathrm{Re} \sum_{\ell m} i \Big(& b_{\ell m}^0 b_{\ell m}^{\mathrm{II}\,*} j_\ell(k_2 R)\xi_\ell'^*(k_2 R) + b_{\ell m}^{0\,*} b_{\ell m}^{\mathrm{II}} \Psi_\ell'^*(k_2 R) h_\ell^+(k_2 R) \\ & - a_{\ell m}^0 a_{\ell m}^{\mathrm{II}\,*} j'_\ell(k_2 R) h_\ell^{+*}(k_2 R) - a_{\ell m}^{0\,*} a_{\ell m}^{\mathrm{II}} j_\ell^*(k_2 R)\xi'_\ell(k_2 R) \Big)\end{aligned} \tag{D.42b}$$

If we now assume an incident L/RCP wave traveling in the $\hat{\mathbf{k}} = \hat{\mathbf{z}}$ direction, then the incident electric field takes the form

$$\mathbf{E}_{\mathrm{inc}} = E_0 \left(\hat{\mathbf{x}} \pm i\hat{\mathbf{y}} \right). \tag{D.43}$$

Substitution of $\mathbf{E}$ and $\hat{\mathbf{k}}$ in the formulas (D.14) giving $a_{\ell m}^0$ and $b_{\ell m}^0$ yields

$$b_{\ell m}^{0;\mathrm{L/R}} = \pm i a_{\ell m}^{0;\mathrm{L/R}}\,. \tag{D.44}$$

We next insert Eq. (D.44) in Eqs. (D.42), and by following similar steps as in the standard Mie theory, we finally obtain

$$\sigma_{\mathrm{scat}}^{\mathrm{L/R}} = \frac{2}{(k_2 R)^2} \sum_\ell (2\ell + 1) \Big[\big|T_{\mathrm{E}\ell}\big|^2 + \big|T_{\mathrm{M}\ell}\big|^2 + 2\big|T_{\mathrm{C}\ell}\big|^2 \mp 2\mathrm{Im}\Big\{ T_{\mathrm{C}\ell}^* \left(T_{\mathrm{E}\ell} + T_{\mathrm{M}\ell}\right) \Big\} \Big]\,, \tag{D.45a}$$

$$\sigma_{\mathrm{ext}}^{\mathrm{L/R}} = -\frac{2}{(k_2 R)^2} \sum_\ell (2\ell + 1) \mathrm{Re}\Big\{ T_{\mathrm{E}\ell} + T_{\mathrm{M}\ell} \mp 2i T_{\mathrm{C}\ell} \Big\}\,. \tag{D.45b}$$

Appendix E
Excitation and Emission of Quantum Emitters Near Nanospheres

In this appendix we show the basic steps of the derivation of the formulas used throughout Chap. 4. As depicted in Fig. 4.1, our system consists of a QE modeled as an electric dipole, and excited by light of frequency ω, in close proximity to a spherical NP characterized by parameters (ε_1, μ_1), embedded in air $(\varepsilon_2, \mu_2 = 1)$.

E.1 Excitation Enhancement

In order to calculate the excitation rate for the various positions and orientations of the QE depicted in Fig. 4.1b, we make use of Eq. (2.108). Then, the problem reduces to calculating the inner product $\mathbf{p}_\mathrm{d} \cdot \mathbf{E}_\mathrm{tot}(\mathbf{r}_\mathrm{d})$, where the total field $\mathbf{E}_\mathrm{tot}$ is the sum of the incident field ($\mathbf{E}_0$) and the one scattered ($\mathbf{E}_\mathrm{scat}$) of the NP surface.

Configuration $\mathbf{A}_1$: In this case we have $\mathbf{p}_\mathrm{d}, \mathbf{r}_\mathrm{d}, \mathbf{E}_0 \parallel \hat{\mathbf{z}}$, and $\mathbf{k} \parallel \hat{\mathbf{x}}$. Therefore,

$$\begin{aligned}
\mathbf{p}_\mathrm{d} \cdot \mathbf{E}_\mathrm{tot}(\mathbf{r}_\mathrm{d}) = p_\mathrm{d}\hat{\mathbf{z}} \cdot \sum_{\ell m} \Big\{ & b^0_{\ell m} j_\ell(k_2 r_\mathrm{d}) \mathbf{X}_{\ell m}(\hat{\mathbf{r}}_\mathrm{d}) + \frac{i}{k_2} a^0_{\ell m} \nabla \times j_\ell(k_2 r_\mathrm{d}) \mathbf{X}_{\ell m}(\hat{\mathbf{r}}_\mathrm{d}) \\
& + b^{\mathrm{II}}_{\ell m} h^+_\ell(k_2 r_\mathrm{d}) \mathbf{X}_{\ell m}(\hat{\mathbf{r}}_\mathrm{d}) + \frac{i}{k_2} a^{\mathrm{II}}_{\ell m} \nabla \times h^+_\ell(k_2 r_\mathrm{d}) \mathbf{X}_{\ell m}(\hat{\mathbf{r}}_\mathrm{d}) \Big\} \\
= p_\mathrm{d} \sum_{\ell m} \Big\{ & -\frac{s_\ell a^0_{\ell m}}{k_2 r_\mathrm{d}} j_\ell(k_2 r_\mathrm{d}) Y^m_\ell(\hat{\mathbf{x}}) + \frac{i a^0_{\ell m}}{k_2 r_\mathrm{d}} \Psi'_\ell(k_2 r_\mathrm{d}) \left[\hat{\mathbf{z}} \times \mathbf{X}_{\ell m}(\hat{\mathbf{z}})\right] \\
& + \frac{b^0_{\ell m}}{s_\ell} j_\ell(k_2 r_\mathrm{d}) \hat{\mathbf{z}} \cdot \Big(\left[\alpha^m_\ell Y^{m+1}_\ell(\hat{\mathbf{z}}) + \alpha^{-m}_\ell Y^{m-1}_\ell(\hat{\mathbf{z}})\right] \hat{\mathbf{x}} \\
& - i \left[\alpha^m_\ell Y^{m+1}_\ell(\hat{\mathbf{z}}) - \alpha^{-m}_\ell Y^{m-1}_\ell(\hat{\mathbf{z}})\right] \hat{\mathbf{y}} + m Y^m_\ell(\hat{\mathbf{z}}) \hat{\mathbf{z}} \Big) \\
& - \frac{s_\ell a^{\mathrm{II}}_{\ell m}}{k_2 r_\mathrm{d}} h^+_\ell(k_2 r_\mathrm{d}) Y^m_\ell(\hat{\mathbf{x}}) + \frac{i a^{\mathrm{II}}_{\ell m}}{k_2 r_\mathrm{d}} \xi'_\ell(k_2 r_\mathrm{d}) \left[\hat{\mathbf{z}} \times \mathbf{X}_{\ell m}(\hat{\mathbf{z}})\right] \\
& + \frac{b^{\mathrm{II}}_{\ell m}}{s_\ell} h^+_\ell(k_2 r_\mathrm{d}) \hat{\mathbf{z}} \cdot \Big(\left[\alpha^m_\ell Y^{m+1}_\ell(\hat{\mathbf{z}}) + \alpha^{-m}_\ell Y^{m-1}_\ell(\hat{\mathbf{z}})\right] \hat{\mathbf{x}}
\end{aligned}$$

P. E. Stamatopoulou, *Strong Light-Matter Interactions in Extreme Plasmonic and Mie-Resonant Systems*, Springer Theses, https://doi.org/10.1007/978-3-031-67907-0

$$
\begin{aligned}
&\qquad - i\left[\alpha_\ell^m Y_\ell^{m+1}(\hat{\mathbf{z}}) - \alpha_\ell^{-m} Y_\ell^{m-1}(\hat{\mathbf{z}})\right]\hat{\mathbf{y}} + m Y_\ell^m(\hat{\mathbf{z}})\hat{\mathbf{z}}\Big)\Big\} \\
&= \frac{p_\mathrm{d}}{\sqrt{4\pi}} \sum_{\ell m} \Big\{ - a_{\ell m}^0 \frac{s_\ell (2\ell+1)}{k_2 r_\mathrm{d}} \delta_{m0} j_\ell(k_2 r_\mathrm{d}) + b_{\ell m}^0 \frac{(2\ell+1)}{s_\ell} m \delta_{m0} j_\ell(k_2 r_\mathrm{d}) \\
&\qquad - a_{\ell m}^{\mathrm{II}} \frac{s_\ell (2\ell+1)}{k_2 r_\mathrm{d}} \delta_{m0} h_\ell^+(k_2 r_\mathrm{d}) + b_{\ell m}^{\mathrm{II}} \frac{(2\ell+1)}{s_\ell} m \delta_{m0} h_\ell^+(k_2 r_\mathrm{d}) \Big\} \\
&= \frac{p_\mathrm{d}}{\sqrt{4\pi}} \sum_{\ell} \Big\{ - a_{\ell 0}^0 \frac{\sqrt{\ell(\ell+1)(2\ell+1)}}{k_2 r_\mathrm{d}} \left[j_\ell(k_2 r_\mathrm{d}) + T_{\mathrm{E}\ell}\, h_\ell^+(k_2 r_\mathrm{d})\right] \Big\}, \qquad \text{(E.1)}
\end{aligned}
$$

where we used Eqs. (C.16), (C.25b) and $s_\ell = \sqrt{\ell(\ell+1)}$. From Eq. (D.14a) we obtain for $a_{\ell 0}^0$

$$
a_{\ell 0}^0 = 2\pi i^{\ell+1} E_0 \left[Y_\ell^{-1}\left(\frac{\pi}{2}, 0\right) - Y_\ell^1\left(\frac{\pi}{2}, 0\right)\right] = 2\pi i^{\ell+1} E_0 \mathscr{Y}_{\ell 1}^- , \qquad \text{(E.2)}
$$

where we introduced

$$
\mathscr{Y}_{\ell j}^{\pm} = Y_\ell^{-j}\left(\frac{\pi}{2}, 0\right) \pm Y_\ell^j\left(\frac{\pi}{2}, 0\right). \qquad \text{(E.3)}
$$

Inserting $a_{\ell 0}^0$ in Eq. (E.1), we find

$$
\begin{aligned}
\mathbf{p}_\mathrm{d} \cdot \mathbf{E}_\mathrm{tot}(\mathbf{r}_\mathrm{d}) &= \frac{p_\mathrm{d} E_0}{\sqrt{4\pi}} \sum_\ell \Big\{ - \frac{2\pi i^{\ell+1} \sqrt{\ell(\ell+1)(2\ell+1)}}{k_2 r_\mathrm{d}} \left[j_\ell(k_2 r_\mathrm{d}) + T_{\mathrm{E}\ell}\, h_\ell^+(k_2 r_\mathrm{d})\right] \mathscr{Y}_{\ell 1}^- \Big\} \Longrightarrow \\
\left|\mathbf{p}_\mathrm{d} \cdot \mathbf{E}_\mathrm{tot}(\mathbf{r}_\mathrm{d})\right|_{\mathrm{A}_1}^2 &= p_\mathrm{d}^2 \pi E_0^2 \left| \sum_\ell \frac{i^{\ell+1} \sqrt{\ell(\ell+1)(2\ell+1)}}{k_2 r_\mathrm{d}} \left[j_\ell(k_2 r_\mathrm{d}) + T_{\mathrm{E}\ell}\, h_\ell^+(k_2 r_\mathrm{d})\right] \mathscr{Y}_{\ell 1}^- \right|^2 .
\end{aligned}
\qquad \text{(E.4)}
$$

Finally, by dividing Eq. (E.4) with the (squared) excitation rate of the QE in free space, $|\mathbf{p}_\mathrm{d} \cdot \mathbf{E}_\mathrm{tot}(\mathbf{r}_\mathrm{d})|^2 = p_\mathrm{d}^2 E_0^2$, we end up with Eq. (4.2a).

Configuration A₂ In this configuration we set $\mathbf{p}_\mathrm{d} \parallel \hat{\mathbf{x}}$, $\mathbf{r}_\mathrm{d}$, $\mathbf{E}_0 \parallel \hat{\mathbf{z}}$, and $\mathbf{k} \parallel \hat{\mathbf{x}}$. Then, we have

$$
\begin{aligned}
\mathbf{p}_\mathrm{d} \cdot \mathbf{E}_\mathrm{tot}(\mathbf{r}_\mathrm{d}) &= p_\mathrm{d} \hat{\mathbf{x}} \cdot \sum_{\ell m} \Big\{ b_{\ell m}^0 j_\ell(k_2 r_\mathrm{d}) \mathbf{X}_{\ell m}(\hat{\mathbf{r}}_\mathrm{d}) + \frac{i}{k_2} a_{\ell m}^0 \nabla \times j_\ell(k_2 r_\mathrm{d}) \mathbf{X}_{\ell m}(\hat{\mathbf{r}}_\mathrm{d}) \\
&\qquad + b_{\ell m}^{\mathrm{II}} h_\ell^+(k_2 r_\mathrm{d}) \mathbf{X}_{\ell m}(\hat{\mathbf{r}}_\mathrm{d}) + \frac{i}{k_2} a_{\ell m}^{\mathrm{II}} \nabla \times h_\ell^+(k_2 r_\mathrm{d}) \mathbf{X}_{\ell m}(\hat{\mathbf{r}}_\mathrm{d}) \Big\} \\
&= p_\mathrm{d} \sum_{\ell m} \Big\{ - \frac{a_{\ell m}^0}{s_\ell k_2 r_\mathrm{d}} \left[\Psi_\ell'(k_2 r_\mathrm{d}) + T_{\mathrm{E}\ell}\, \xi_\ell'(k_2 r_\mathrm{d})\right] \left[\alpha_\ell^m Y_\ell^{m+1}(\hat{\mathbf{z}}) - \alpha_\ell^{-m} Y_\ell^{m-1}(\hat{\mathbf{z}})\right] \\
&\qquad + \frac{b_{\ell m}^0}{s_\ell} \left[j_\ell(k_2 r_\mathrm{d}) + T_{\mathrm{M}\ell}\, h_\ell^+(k_2 r_\mathrm{d})\right] \left[\alpha_\ell^m Y_\ell^{m+1}(\hat{\mathbf{z}}) + \alpha_\ell^{-m} Y_\ell^{m-1}(\hat{\mathbf{z}})\right] \Big\} \\
&= \frac{p_\mathrm{d}}{\sqrt{4\pi}} \sum_{\ell m} \Big\{ - \frac{(2\ell+1) a_{\ell m}^0}{s_\ell k_2 r_\mathrm{d}} \left[\Psi_\ell'(k_2 r_\mathrm{d}) + T_{\mathrm{E}\ell}\, \xi_\ell'(k_2 r_\mathrm{d})\right] \left[\alpha_\ell^m \delta_{m-1} - \alpha_\ell^{-m} \delta_{m1}\right] \\
&\qquad + \frac{(2\ell+1) b_{\ell m}^0}{s_\ell} \left[j_\ell(k_2 r_\mathrm{d}) + T_{\mathrm{M}\ell}\, h_\ell^+(k_2 r_\mathrm{d})\right] \left[\alpha_\ell^m \delta_{m-1} + \alpha_\ell^{-m} \delta_{m1}\right] \Big\}
\end{aligned}
$$

$$= \frac{p_\mathrm{d}}{\sqrt{4\pi}} \sum_\ell \sqrt{2\ell+1} \Bigg\{ \frac{1}{k_2 r_\mathrm{d}} \left[\Psi'_\ell(k_2 r_\mathrm{d}) + T_{\mathrm{E}\ell}\, \xi'_\ell(k_2 r_\mathrm{d})\right] \left[a^0_{\ell 1} - a^0_{\ell -1}\right]$$
$$+ \left[j_\ell(k_2 r_\mathrm{d}) + T_{\mathrm{M}\ell}\, h^+_\ell(k_2 r_\mathrm{d})\right] \left[b^0_{\ell -1} + b^0_{\ell 1}\right] \Bigg\} . \tag{E.5}$$

Using Eqs. (D.14) we obtain

$$a^0_{\ell 1} - a^0_{\ell -1} = 2\pi i^{\ell+1} E_0 \mathscr{Y}^+_{\ell 0} - 2\pi i^{\ell+1} E_0 \sqrt{\frac{(\ell-1)(\ell+2)}{\ell(\ell+1)}} \mathscr{Y}^+_{\ell 2} , \tag{E.6a}$$

$$b^0_{\ell 1} - b^0_{\ell -1} = -\frac{4\pi i^\ell E_0}{s_\ell} \mathscr{Y}^-_{\ell 1} . \tag{E.6b}$$

Inserting Eqs. (E.6) into (E.5), we end up with

$$\left|\mathbf{p}_\mathrm{d} \cdot \mathbf{E}_\mathrm{tot}(\mathbf{r}_\mathrm{d})\right|^2_{\mathrm{A}_2} = \frac{p_\mathrm{d}^2 \pi E_0^2}{4} \Bigg| \sum_\ell \Bigg\{ -2 i^\ell \sqrt{\frac{2\ell+1}{\ell(\ell+1)}} \left[j_\ell(k_2 r_\mathrm{d}) + T_{\mathrm{M}\ell}\, h^+_\ell(k_2 r_\mathrm{d})\right] \mathscr{Y}^-_{\ell 1}$$
$$+ \frac{i^{\ell+1}\sqrt{2\ell+1}}{k_2 r_\mathrm{d}} \left[\Psi'_\ell(k_2 r_\mathrm{d}) + T_{\mathrm{E}\ell}\, \xi'_\ell(k_2 r_\mathrm{d})\right] \left[\mathscr{Y}^+_{\ell 0} - \sqrt{\frac{(\ell-1)(\ell+2)}{\ell(\ell+1)}} \mathscr{Y}^+_{\ell 2}\right] \Bigg\} \Bigg|^2 . \tag{E.7}$$

Normalization of Eq. (E.1) with $p_\mathrm{d}^2 E_0^2$ yields Eq. (4.2b).

Configuration B In this case we set $\mathbf{p}_\mathrm{d}, \mathbf{E}_0 \parallel \hat{\mathbf{z}}$, $\mathbf{r}_\mathrm{d} \parallel \hat{\mathbf{y}}$, and $\mathbf{k} \parallel \hat{\mathbf{x}}$. If we rotate the coordinate system, by $\pi/2$ around the x axis, we have $\mathbf{r}_\mathrm{d} \parallel \hat{\mathbf{z}}$, $\mathbf{E}_0 \parallel -\hat{\mathbf{y}}$, $\mathbf{p}_\mathrm{d} \parallel -\hat{\mathbf{y}}$, and $\mathbf{k} \parallel \hat{\mathbf{x}}$. In the new coordinate system we find

$$\mathbf{p}_\mathrm{d} \cdot \mathbf{E}_\mathrm{tot}(\mathbf{r}_\mathrm{d}) = -p_\mathrm{d} \hat{\mathbf{y}} \cdot \sum_{\ell m} \Big\{ b^0_{\ell m} j_\ell(k_2 r_\mathrm{d}) \mathbf{X}_{\ell m}(\hat{\mathbf{r}}_\mathrm{d}) + \frac{i}{k_2} a^0_{\ell m} \nabla \times j_\ell(k_2 r_\mathrm{d}) \mathbf{X}_{\ell m}(\hat{\mathbf{r}}_\mathrm{d})$$
$$+ b^{\mathrm{II}}_{\ell m} h^+_\ell(k_2 r_\mathrm{d}) \mathbf{X}_{\ell m}(\hat{\mathbf{r}}_\mathrm{d}) + \frac{i}{k_2} a^{\mathrm{II}}_{\ell m} \nabla \times h^+_\ell(k_2 r_\mathrm{d}) \mathbf{X}_{\ell m}(\hat{\mathbf{r}}_\mathrm{d}) \Big\}$$
$$= -p_\mathrm{d} \sum_{\ell m} \Bigg\{ \frac{i a^0_{\ell m}}{s_\ell k_2 r_\mathrm{d}} \left[\Psi'_\ell(k_2 r_\mathrm{d}) + T_{\mathrm{E}\ell}\, \xi'_\ell(k_2 r_\mathrm{d})\right] \hat{\mathbf{y}} \cdot \Big(\hat{\mathbf{z}} \times \Big\{ [\alpha^m_\ell Y^{m+1}_\ell(\hat{\mathbf{z}}) + \alpha^{-m}_\ell Y^{m-1}_\ell(\hat{\mathbf{z}})]\hat{\mathbf{x}}$$
$$- i[\alpha^m_\ell Y^{m+1}_\ell(\hat{\mathbf{z}}) - \alpha^{-m}_\ell Y^{m-1}_\ell(\hat{\mathbf{z}})]\hat{\mathbf{y}} + m Y^m_\ell(\hat{\mathbf{z}})\hat{\mathbf{z}} \Big\} \Big)$$
$$- \frac{i b^0_{\ell m}}{s_\ell} \left[j_\ell(k_2 r_\mathrm{d}) + T_{\mathrm{M}\ell}\, h^+_\ell(k_2 r_\mathrm{d})\right] \left[\alpha^m_\ell Y^{m+1}_\ell(\hat{\mathbf{z}}) - \alpha^{-m}_\ell Y^{m-1}_\ell(\hat{\mathbf{z}})\right] \Bigg\}$$
$$= -\frac{p_\mathrm{d}}{\sqrt{4\pi}} \sum_{\ell m} \Bigg\{ \frac{i a^0_{\ell m}}{k_2 r_\mathrm{d}} \left[\Psi'_\ell(k_2 r_\mathrm{d}) + T_{\mathrm{E}\ell}\, \xi'_\ell(k_2 r_\mathrm{d})\right] \sqrt{\frac{2\ell+1}{\ell(\ell+1)}} \left[\alpha^m_\ell \delta_{m-1} + \alpha^{-m}_\ell \delta_{m1}\right]$$
$$- i b^0_{\ell m} \left[j_\ell(k_2 r_\mathrm{d}) + T_{\mathrm{M}\ell}\, h^+_\ell(k_2 r_\mathrm{d})\right] \sqrt{\frac{2\ell+1}{\ell(\ell+1)}} \left[\alpha^m_\ell \delta_{m-1} - \alpha^{-m}_\ell \delta_{m1}\right] \Bigg\}$$

$$= -\frac{ip_{\rm d}}{4\sqrt{\pi}} \sum_\ell \sqrt{2\ell+1} \left\{ \frac{\Psi'_\ell(k_2 r_{\rm d}) + T_{{\rm E}\ell}\, \xi'_\ell(k_2 r_{\rm d})}{k_2 r_{\rm d}} \left[a^0_{\ell -1} + a^0_{\ell 1} \right] \right.$$
$$\left. - \left[j_\ell(k_2 r_{\rm d}) + T_{{\rm M}\ell}\, h^+_\ell(k_2 r_{\rm d}) \right] \left[b^0_{\ell -1} - b^0_{\ell 1} \right] \right\}. \tag{E.8}$$

Recalling that, after the rotation of the coordinate system, $\mathbf{E}_0$ is along the negative y axis, we must calculate anew $a^0_{\ell \pm 1}$ and $b^0_{\ell \pm 1}$:

$$a^0_{\ell -1} + a^0_{\ell 1} = \frac{4\pi i^\ell E_0}{s_\ell} \mathscr{Y}^-_{\ell 1}, \tag{E.9a}$$

$$b^0_{\ell -1} - b^0_{\ell 1} = -2\pi i^{\ell+1} E_0 \mathscr{Y}^+_{\ell 0} + 2\pi i^{\ell+1} E_0 \sqrt{\frac{(\ell-1)(\ell+2)}{\ell(\ell+1)}} \mathscr{Y}^+_{\ell 2}. \tag{E.9b}$$

Consequently, we obtain

$$\left| \mathbf{p}_{\rm d} \cdot \mathbf{E}_{\rm tot}(\mathbf{r}_{\rm d}) \right|^2_{\rm B} = \frac{p^2_{\rm d} \pi E^2_0}{4} \left| \sum_\ell \frac{2 i^\ell}{k_2 r_{\rm d}} \sqrt{\frac{2\ell+1}{\ell(\ell+1)}} \left[\Psi'_\ell(k_2 r_{\rm d}) + T_{{\rm E}\ell}\, \xi'_\ell(k_2 r_{\rm d}) \right] \mathscr{Y}^-_{\ell 1} \right.$$
$$\left. + i^{\ell+1} \sqrt{2\ell+1} \left[j_\ell(k_2 r_{\rm d}) + T_{{\rm M}\ell}\, h^+_\ell(k_2 r_{\rm d}) \right] \left[\mathscr{Y}^+_{\ell 0} - \sqrt{\frac{(\ell-1)(\ell+2)}{\ell(\ell+1)}} \mathscr{Y}^+_{\ell 2} \right] \right|^2, \tag{E.10}$$

which leads to Eq. (4.2c) once divided with $p^2_{\rm d} E^2_0$.

Configuration C Here we have $\mathbf{p}_{\rm d}$, $\mathbf{E}_0 \parallel \hat{\mathbf{z}}$ and $\mathbf{r}_{\rm d}$, $\mathbf{k} \parallel \hat{\mathbf{x}}$. We again proceed with a rotation of the coordinate system, so that in the new system $\mathbf{r}_{\rm d} \parallel \hat{\mathbf{z}}$, $\mathbf{E} \parallel -\hat{\mathbf{x}}$, $\mathbf{p}_{\rm d} \parallel -\hat{\mathbf{x}}$, and $\mathbf{k} \parallel \hat{\mathbf{z}}$. Then we obtain

$$\mathbf{p}_{\rm d} \cdot \mathbf{E}_{\rm tot}(\mathbf{r}_{\rm d}) = -p_{\rm d} \hat{\mathbf{x}} \cdot \sum_{\ell m} \left\{ b^0_{\ell m} j_\ell(k_2 r_{\rm d}) \mathbf{X}_{\ell m}(\hat{\mathbf{r}}_{\rm d}) + \frac{i}{k_2} a^0_{\ell m} \nabla \times j_\ell(k_2 r_{\rm d}) \mathbf{X}_{\ell m}(\hat{\mathbf{r}}_{\rm d}) \right.$$
$$\left. + b^{\rm II}_{\ell m} h^+_\ell(k_2 r_{\rm d}) \mathbf{X}_{\ell m}(\hat{\mathbf{r}}_{\rm d}) + \frac{i}{k_2} a^{\rm II}_{\ell m} \nabla \times h^+_\ell(k_2 r_{\rm d}) \mathbf{X}_{\ell m}(\hat{\mathbf{r}}_{\rm d}) \right\}$$
$$= -p_{\rm d} \sum_{\ell m} \left\{ \frac{i a^0_{\ell m}}{k_2 r_{\rm d}} \left[\Psi'_\ell(k_2 r_{\rm d}) + T_{{\rm E}\ell}\, \xi'_\ell(k_2 r_{\rm d}) \right] \hat{\mathbf{x}} \cdot \left[\hat{\mathbf{z}} \times \mathbf{X}_{\ell m}(\hat{\mathbf{z}}) \right] \right.$$
$$+ \frac{b^0_{\ell m}}{s_\ell} \left[j_\ell(k_2 r_{\rm d}) + T_{{\rm M}\ell}\, h^+_\ell(k_2 r_{\rm d}) \right] \hat{\mathbf{x}} \cdot \left(\left[\alpha^m_\ell Y^{m+1}_\ell(\hat{\mathbf{z}}) + \alpha^{-m}_\ell Y^{m-1}_\ell(\hat{\mathbf{z}}) \right] \hat{\mathbf{x}} \right.$$
$$\left. \left. - i \left[\alpha^m_\ell Y^{m+1}_\ell(\hat{\mathbf{z}}) - \alpha^{-m}_\ell Y^{m-1}_\ell(\hat{\mathbf{z}}) \right] \hat{\mathbf{y}} + m Y^m_\ell(\hat{\mathbf{z}}) \hat{\mathbf{z}} \right) \right\}$$
$$= -\frac{p_{\rm d}}{\sqrt{4\pi}} \sum_{\ell m} \left\{ -\frac{a^0_{\ell m}}{k_2 r_{\rm d}} \left[\Psi'_\ell(k_2 r_{\rm d}) + T_{{\rm E}\ell}\, \xi'_\ell(k_2 r_{\rm d}) \right] \sqrt{\frac{2\ell+1}{\ell(\ell+1)}} \left[\alpha^m_\ell \delta_{m-1} - \alpha^{-m}_\ell \delta_{m1} \right] \right.$$

$$+ b^0_{\ell m}\left[j_\ell(k_2r_\mathrm{d}) + T_{\mathrm{M}\ell}\, h^+_\ell(k_2r_\mathrm{d})\right]\sqrt{\frac{2\ell+1}{\ell(\ell+1)}}\left[\alpha^m_\ell\delta_{m-1} + \alpha^{-m}_\ell\delta_{m1}\right]\Bigg\}$$

$$= \frac{p_\mathrm{d}}{4\sqrt{\pi}}\sum_\ell\sqrt{2\ell+1}\Bigg\{\frac{\Psi'_\ell(k_2r_\mathrm{d}) + T_{\mathrm{E}\ell}\,\xi'_\ell(k_2r_\mathrm{d})}{k_2r_\mathrm{d}}\left[a^0_{\ell-1} - a^0_{\ell 1}\right]$$

$$-\left[j_\ell(k_2r_\mathrm{d}) + T_{\mathrm{M}\ell}\, h^+_\ell(k_2r_\mathrm{d})\right]\left[b^0_{\ell-1} + b^0_{\ell 1}\right]\Bigg\}. \tag{E.11}$$

In this case, $\mathbf{E}_0$ is along the negative x axis, therefore

$$a^0_{\ell-1} - a^0_{\ell 1} = -2\sqrt{\pi}i^{\ell+1}E_0\sqrt{2\ell+1}\,, \tag{E.12a}$$

$$b^0_{\ell-1} + b^0_{\ell 1} = -2\sqrt{\pi}i^{\ell}E_0\sqrt{2\ell+1}\,. \tag{E.12b}$$

Inserting these relations in Eq. (E.11) we obtain

$$\left|\mathbf{p}_\mathrm{d}\cdot\mathbf{E}_\mathrm{tot}(\mathbf{r}_\mathrm{d})\right|^2_\mathrm{C} = \frac{p_\mathrm{d}^2E_0^2}{4}\Bigg|\sum_\ell(2\ell+1)\Bigg\{\frac{i^{\ell+1}}{k_2r_\mathrm{d}}\left[\Psi'_\ell(k_2r_\mathrm{d}) + T_{\mathrm{E}\ell}\,\xi'_\ell(k_2r_\mathrm{d})\right]$$

$$- i^\ell\left[j_\ell(k_2r_\mathrm{d}) + T_{\mathrm{M}\ell}\,h^+_\ell(k_2r_\mathrm{d})\right]\Bigg\}\Bigg|^2. \tag{E.13}$$

By dividing with $p_\mathrm{d}^2E_0^2$, we end up with Eq. (4.2d).

Configuration D Finally, configuration D is the antipodal point of C. The excitation rate is the result of Eq. (E.11), multiplied by $(-1)^\ell$, resulting in

$$\left|\mathbf{p}_\mathrm{d}\cdot\mathbf{E}_\mathrm{tot}(\mathbf{r}_\mathrm{d})\right|^2_\mathrm{D} = \frac{p_\mathrm{d}^2E_0^2}{4}\Bigg|\sum_\ell(-1)^\ell(2\ell+1)\Bigg\{\frac{i^{\ell+1}}{k_2r_\mathrm{d}}\left[\Psi'_\ell(k_2r_\mathrm{d}) + T_{\mathrm{E}\ell}\,\xi'_\ell(k_2r_\mathrm{d})\right]$$

$$- i^\ell\left[j_\ell(k_2r_\mathrm{d}) + T_{\mathrm{M}\ell}\,h^+_\ell(k_2r_\mathrm{d})\right]\Bigg\}\Bigg|^2. \tag{E.14}$$

E.2 Photonic Enhancement

Once the QE is excited we “switch off” the incident plane wave and let the NP–QE system decay. Recalling Eq. (2.106), the electric field generated by the excited QE is that of an oscillating point dipole

$$\mathbf{E}(\mathbf{r}) = \omega^2\mu_2\mu_0\overset{\leftrightarrow}{\mathscr{G}}_0(\mathbf{r}, \mathbf{r}_\mathrm{d}, \omega)\cdot\mathbf{p}_\mathrm{d}\,. \tag{E.15}$$

Expanded on the basis of spherical waves, the QE field takes the form of incoming spherical waves for $r < r_\mathrm{d}$ and outgoing everywhere else

$$\mathbf{E}(\mathbf{r}) = \sum_{\ell m} \Big\{ \Big[d^{0}_{\mathrm{H};\ell m} j_\ell(k_2 r)\mathbf{X}_{\ell m}(\hat{\mathbf{r}}) + \frac{i}{k_2} d^{0}_{\mathrm{E};\ell m} \nabla \times j_\ell(k_2 r)\mathbf{X}_{\ell m}(\hat{\mathbf{r}}) \Big] \Theta(r_\mathrm{d} - r)$$
$$+ \Big[d^{+}_{\mathrm{H};\ell m} h^{+}_\ell(k_2 r)\mathbf{X}_{\ell m}(\hat{\mathbf{r}}) + \frac{i}{k_2} d^{+}_{\mathrm{E};\ell m} \nabla \times h^{+}_\ell(k_2 r)\mathbf{X}_{\ell m}(\hat{\mathbf{r}}) \Big] \Theta(r - r_\mathrm{d}) \Big\} , \tag{E.16}$$

where $\Theta(x)$ is the Heaviside step function [2]. According to Eq. (E.15), the expansion coefficients of the dipole field read

$$d^{0}_{\mathrm{E};\ell m} = \frac{i k_2^3 p_\mathrm{d}}{\varepsilon_2 \varepsilon_0} \mathbf{p}_\mathrm{d} \cdot \frac{i}{k_2} \nabla \times h^{+}_\ell(k_2 r)\mathbf{X}^{*}_{\ell m}(k_2 r) , \tag{E.17a}$$

$$d^{0}_{\mathrm{M};\ell m} = \frac{i k_2^3 p_\mathrm{d}}{\varepsilon_2 \varepsilon_0} \mathbf{p}_\mathrm{d} \cdot h^{+}_\ell(k_2 r)\mathbf{X}^{*}_{\ell m}(k_2 r) , \tag{E.17b}$$

$$d^{+}_{\mathrm{E};\ell m} = \frac{i k_2^3 p_\mathrm{d}}{\varepsilon_2 \varepsilon_0} \mathbf{p}_\mathrm{d} \cdot \frac{i}{k_2} \nabla \times j_\ell(k_2 r)\mathbf{X}^{*}_{\ell m}(k_2 r) , \tag{E.17c}$$

$$d^{+}_{\mathrm{M};\ell m} = \frac{i k_2^3 p_\mathrm{d}}{\varepsilon_2 \varepsilon_0} \mathbf{p}_\mathrm{d} \cdot j_\ell(k_2 r)\mathbf{X}^{*}_{\ell m}(k_2 r) . \tag{E.17d}$$

For $r < r_\mathrm{d}$ the problem is identical to traditional Mie scattering (see Appendix D), albeit for the incident plane wave of Eqs. (D.13) replaced by the QE field, as $a^0_{\ell m} \to d^0_{\mathrm{E};\ell m}$, and $b^0_{\ell m} \to d^0_{\mathrm{M};\ell m}$. The internal field of the NP, as well as the scattered, admit once again the form of Eqs. (D.9) and (D.11), respectively. Therefore the boundary-condition problem remains the same, which allows us to utilize the Mie coefficients (D.17) and (D.19).

Radiative Photonic Enhancement In order to obtain the total scattered energy, one needs to calculate the energy flow through a sphere of radius R that encloses the NP–QE system. To calculate the power radiated we can make use of Eq. (D.21a), taking into account that both the NP and the QE give rise to outgoing spherical waves. This is readily done via the replacements: $a^{\mathrm{II}}_{\ell m} \to a^{\mathrm{II}}_{\ell m} + d^{+}_{\mathrm{E};\ell m}$, and $b^{\mathrm{II}}_{\ell m} \to b^{\mathrm{II}}_{\ell m} + d^{+}_{\mathrm{M};\ell m}$, i.e.

$$P_\mathrm{r} = \frac{1}{2k_2^2 Z_2} \sum_{\ell m} \Big\{ \big| a^{\mathrm{II}}_{\ell m} + d^{+}_{\mathrm{E};\ell m} \big|^2 + \big| b^{\mathrm{II}}_{\ell m} + d^{+}_{\mathrm{M};\ell m} \big|^2 \Big\}$$
$$= \frac{1}{2k_2^2 Z_2} \sum_{\ell m} \Big\{ \big| T_{\mathrm{E}\ell}\, d^{0}_{\mathrm{E};\ell m} + d^{+}_{\mathrm{E};\ell m} \big|^2 + \big| T_{\mathrm{M}\ell}\, d^{0}_{\mathrm{M};\ell m} + d^{+}_{\mathrm{M};\ell m} \big|^2 \Big\} . \tag{E.18}$$

After some lengthy algebra involving sum rules for the vector spherical harmonics, we end up with

$$P_\mathrm{r} = \frac{k_2^2}{2\varepsilon_0^2 \varepsilon_2^2 r_\mathrm{d}^2} \sqrt{\frac{\varepsilon_0 \varepsilon_2}{\mu_0 \mu_2}} \sum_\ell \Big\{ \frac{\ell(\ell+1)(2\ell+1)}{4\pi} \big| T_{\mathrm{E}\ell}\, h^{+}_\ell(k_2 r_\mathrm{d}) + j_\ell(k_2 r_\mathrm{d}) \big|^2 p_r^2$$

$$+\frac{2\ell+1}{8\pi}\left|T_{\mathrm{E}\ell}\,\xi_\ell'(k_2 r_\mathrm{d})+\Psi_\ell'(k_2 r_\mathrm{d})\right|^2\left(p_\theta^2+p_\phi^2\right)\Bigg\} \tag{E.19}$$

$$+\frac{k_2^4}{2\varepsilon_0^2\varepsilon_2^2}\sqrt{\frac{\varepsilon_0\varepsilon_2}{\mu_0\mu_2}}\sum_\ell\left\{\frac{2\ell+1}{8\pi}\left|T_{\mathrm{H}\ell}\,h_\ell^+(k_2 r_\mathrm{d})+j_\ell(k_2 r_\mathrm{d})\right|^2\left(p_\theta^2+p_\phi^2\right)\right\}, \tag{E.20}$$

where p_r, p_θ, and p_ϕ are the three components of $\mathbf{p}_\mathrm{d}$ in spherical coordinates. If the QE oscillates along the direction connecting the center of the NP with the QE (case $\|$), then $p_\mathrm{d}=p_r$ and $p_\theta=p_\phi=0$. In this case, we find

$$\frac{P_\mathrm{r}^{\|}}{P_0}=\frac{3}{2}\frac{1}{k_2^2 r_\mathrm{d}^2}\sum_\ell \ell\,(\ell+1)\,(2\ell+1)\left|j_\ell(k_2 r_\mathrm{d})+T_{\mathrm{E}\ell}h_\ell^+(k_2 r_\mathrm{d})\right|^2. \tag{E.21}$$

Similarly, for a dipole oscillating perpendicularly to the direction connecting the center of the NP with the QE (case $\perp$), then $p_r=0$ and $p_\theta^2+p_\phi^2=p_\mathrm{d}^2$, yielding

$$\frac{P_\mathrm{r}^{\perp}}{P_0}=\frac{3}{4}\sum_\ell(2\ell+1)\left\{\frac{1}{k_2^2 r_\mathrm{d}^2}\left|\Psi_\ell'(k_2 r_\mathrm{d})+T_{\mathrm{E}\ell}\xi'(k_2 r_\mathrm{d})\right|^2+\left|j_\ell(k_2 r_\mathrm{d})+T_{\mathrm{M}\ell}h_\ell^+(k_2 r_\mathrm{d})\right|^2\right\}. \tag{E.22}$$

Non-radiative Photonic Enhancement From Eq. (D.21b), we have

$$\begin{aligned}P_\mathrm{nr}&=-\frac{1}{2k_2^2 Z_2}\sum_{\ell m}\Big(\left|a_{\ell m}^{\mathrm{II}}+d_{\mathrm{E};\ell m}^+\right|^2+\left|b_{\ell m}^{\mathrm{II}}+d_{\mathrm{M};\ell m}^+\right|^2\\&\qquad+\mathrm{Re}\Big\{T_{\mathrm{E}\ell}\left|d_{\mathrm{E};\ell m}^0\right|^2+T_{\mathrm{M}\ell}\left|d_{\mathrm{M};\ell m}^0\right|^2\Big\}\Big)\\&=-\frac{1}{2k_2^2 Z_2}\sum_{\ell m}\Big(\left|T_{\mathrm{E}\ell}\,d_{\mathrm{E};\ell m}^0+d_{\mathrm{E};\ell m}^+\right|^2+\left|T_{\mathrm{M}\ell}\,d_{\mathrm{M};\ell m}^0+d_{\mathrm{M};\ell m}^+\right|^2\\&\qquad+\mathrm{Re}\Big\{T_{\mathrm{E}\ell}\left|d_{\mathrm{E};\ell m}^0\right|^2+T_{\mathrm{M}\ell}\left|d_{\mathrm{M};\ell m}^0\right|^2\Big\}\Big).\end{aligned} \tag{E.23}$$

Similarly to the case of radiative losses, an analysis based on sum rules for the vector spherical harmonics eventually yields

$$\begin{aligned}\frac{P_\mathrm{nr}}{P_0}=&-\frac{3}{2}\frac{1}{k_2^2 r_\mathrm{d}^2 p_\mathrm{d}^2}\sum_\ell\Big\{\left[\mathrm{Re}\{T_{\mathrm{E}\ell}\}+|T_{\mathrm{E}\ell}|^2\right]\ell\,(\ell+1)\,(2\ell+1)\left|h_\ell^+(k_2 r_\mathrm{d})\right|^2 p_r^2\Big\}\\&-\frac{3}{4}\frac{1}{k_2^2 r_\mathrm{d}^2 p_\mathrm{d}^2}\sum_\ell\Big\{\left[\mathrm{Re}\{T_{\mathrm{E}\ell}\}+|T_{\mathrm{E}\ell}|^2\right](2\ell+1)\left|\xi_\ell'(k_2 r_\mathrm{d})\right|^2\left(p_\theta^2+p_\phi^2\right)\Big\}\\&-\frac{3}{4p_\mathrm{d}^2}\sum_\ell\Big\{\left[\mathrm{Re}\{T_{\mathrm{M}\ell}\}+|T_{\mathrm{M}\ell}|^2\right](2\ell+1)\left|h_\ell^+(k_2 r_\mathrm{d})\right|^2\left(p_\theta^2+p_\phi^2\right)\Big\}.\end{aligned} \tag{E.24}$$

Therefore, in the special case of a dipole oscillating in the radial direction ($p_d = p_r$) we have

$$\frac{P_{\mathrm{nr}}^{\parallel}}{P_0} = -\frac{3}{2(k_2 r_{\mathrm{d}})^2} \sum_{\ell} \ell(\ell+1)(2\ell+1)\Big[\mathrm{Re}\{T_{\mathrm{E}\ell}\} + |T_{\mathrm{E}\ell}|^2\Big] \big|h_\ell^+(k_2 r_{\mathrm{d}})\big|^2 , \quad \text{(E.25)}$$

while for a dipole in the tangential direction we have

$$\frac{P_{\mathrm{nr}}^{\perp}}{P_0} = -\frac{3}{4} \sum_{\ell} (2\ell+1) \Bigg\{ \Big[\mathrm{Re}\{T_{\mathrm{E}\ell}\} + |T_{\mathrm{E}\ell}|^2\Big] \left|\frac{\xi_\ell'(k_2 r_{\mathrm{d}})}{k_2 r_{\mathrm{d}}}\right|^2 + \Big[\mathrm{Re}\{T_{\mathrm{M}\ell}\} + |T_{\mathrm{M}\ell}|^2\Big] \big|h_\ell^+(k_2 r_{\mathrm{d}})\big|^2 \Bigg\} . \quad \text{(E.26)}$$

Appendix F
Electron-Beam Spectroscopy of Nanospheres

F.1 Electromagnetic Field of a Swift Electron

In what follows we provide a detailed derivation of the field of a swift electron traveling inside the bulk of an arbitrary isotropic non-absorbing medium (ε, μ), with constant velocity $\mathbf{v}$.[4] The field of a moving point charge can be found in most standard electrodynamics textbooks, for instance Refs. [4, 10]. Here, we follow the work of García de Abajo [11, 12], where the field is decomposed into a TE and a TM component, both of which are expanded on the basis of spherical waves.

Without loss of generality, we assume an electron traveling on the z axis ($\mathbf{v} = v\hat{\mathbf{z}}$), along a straight trajectory $\mathbf{r}_e = \mathbf{r}_0 + \mathbf{v}t$, where $r_0 = (x_0, y_0, z_0)$ denotes its initial position. The electron is treated as a point charge $-e$, characterized by charge density $\rho_f(\mathbf{r}, t) = -e\delta[\mathbf{r} - \mathbf{r}_e(t)]$ and current density $\mathbf{J}_f(\mathbf{r}, t) = \rho_f(\mathbf{r}, t)\mathbf{v}$. Invoking similar arguments as in Appendix D, the electric field can be decomposed into TE and TM waves

$$\mathbf{E} = \mathbf{L}\psi^{\mathrm{M}} + \frac{i}{k}\nabla \times \mathbf{L}\psi^{\mathrm{E}} , \tag{F.1}$$

where the scalar functions

$$\psi^{\mathrm{M}} = \frac{1}{\mathrm{L}^2}\mathbf{L} \cdot \mathbf{E} , \quad \text{and} \quad \psi^{\mathrm{E}} = \frac{ik}{\mathrm{L}^2\nabla^2}(\mathbf{L} \times \nabla) \cdot \mathbf{E} \tag{F.2}$$

satisfy the Helmholtz equation $\left(\nabla^2 + k^2\right)\psi = 0$, with $k = \omega\sqrt{\varepsilon\mu}/c$ indicating the wave number inside the medium. Therefore, the electric field acquires the general form (D.6)

---

[4] Throughout the derivation we tacitly assume that $v\sqrt{\varepsilon\mu} < c$, which allows us to make use of Eq. (F.31). In principle, one may extend the calculation for $v\sqrt{\varepsilon\mu} > c$ as well, but since in the examples considered here air is always the host medium, such a calculation is not really of interest.

P. E. Stamatopoulou, *Strong Light-Matter Interactions in Extreme Plasmonic and Mie-Resonant Systems*, Springer Theses, https://doi.org/10.1007/978-3-031-67907-0

$$\mathbf{E}(\mathbf{r}) = \sum_{\ell m} \left\{ b^0_{\ell m} j_\ell(kr) \mathbf{X}_{\ell m}(\hat{\mathbf{r}}) + \frac{i}{k} a^0_{\ell m} \nabla \times j_\ell(kr) \mathbf{X}_{\ell m}(\hat{\mathbf{r}}) \right\} , \tag{F.3}$$

where we chose to expand the field as standing waves $f_\ell = j_\ell$ for reasons that become apparent later on.

Finding the field of the electron reduces to finding the expansion coefficients $a^0_{\ell m}$ and $b^0_{\ell m}$. In order to do so, we solve the set of scalar and vector potential Helmholtz equations (2.29), that read

$$\left(\nabla^2 + k^2\right)\varphi(\mathbf{r}, t) = \frac{e}{\varepsilon\varepsilon_0}\delta[\mathbf{r} - \mathbf{r}_e(t)] , \tag{F.4a}$$

$$\left(\nabla^2 + k^2\right)\mathbf{A}(\mathbf{r}, t) = e\mu\mu_0\mathbf{v}\delta[\mathbf{r} - \mathbf{r}_e(t)] . \tag{F.4b}$$

It is, then, easy to recognize the emergence of the Green's function $\mathscr{G}$ for the Helmholtz equation in spherical coordinates

$$\left(\nabla^2 + k^2\right)\mathscr{G}(\mathbf{r} - \mathbf{r}') = -\delta(\mathbf{r} - \mathbf{r}') . \tag{F.5}$$

Following the conventions utilized in Ref. [4], the representation of $\mathscr{G}$ in spherical waves reads

$$\mathscr{G}(\mathbf{r} - \mathbf{r}') = \frac{\mathrm{e}^{ik|\mathbf{r}-\mathbf{r}'|}}{4\pi|\mathbf{r} - \mathbf{r}'|} = ik \sum_{\ell=0}^{\infty} j_\ell(kr_<) h^+_\ell(kr_>) \sum_{m=-\ell}^{\ell} Y^m_\ell(\hat{\mathbf{r}}) Y^{m*}_\ell(\hat{\mathbf{r}}') , \tag{F.6}$$

where $r_{</>}$ stands for the smaller/greater between r and r'. Then the potentials take the form

$$\varphi(\mathbf{r}, t) = -\frac{e}{\varepsilon\varepsilon_0}\mathscr{G}[\mathbf{r} - \mathbf{r}_e(t)] , \tag{F.7a}$$

$$\mathbf{A}(\mathbf{r}, t) = -e\mu\mu_0\mathbf{v}\mathscr{G}[\mathbf{r} - \mathbf{r}_e(t)] , \tag{F.7b}$$

and the electric field, according to Eq. (2.27) reads

$$\mathbf{E}(\mathbf{r}, t) = \frac{e}{\varepsilon\varepsilon_0}\left\{\nabla - \frac{ik^2}{\omega}\mathbf{v}\right\}\mathscr{G}[\mathbf{r} - \mathbf{r}_e(t)] , \tag{F.8}$$

or, in the frequency domain,

$$\mathbf{E}(\mathbf{r}, \omega) = \frac{e}{\varepsilon\varepsilon_0}\left\{\nabla - \frac{ik^2}{\omega}\mathbf{v}\right\} \int_{-\infty}^{+\infty} dt\, \mathrm{e}^{i\omega t}\mathscr{G}[\mathbf{r} - \mathbf{r}_e(t)] . \tag{F.9}$$

Assuming $r_e(t) > r$,[5] the electric field can be expanded to a combination of standing waves

[5] We are free to set the axes origin wherever we wish.

$$\mathbf{E}(\mathbf{r},\omega)=\frac{e}{\varepsilon\varepsilon_0}\left\{\nabla-\frac{ik^2}{\omega}\mathbf{v}\right\}\sum_{\ell m}\phi^{+}_{\ell m}j_\ell(kr)Y_\ell^m(\hat{\mathbf{r}}),\tag{F.10}$$

where the 0-order term of the sum can be omitted, since it does not contribute. In Eq. (F.10) we have set

$$\phi^{\pm}_{\ell m}=ik\int_{-\infty}^{+\infty}dt\,\mathrm{e}^{\pm i\omega t}h_\ell^{+}(kr_e)Y_\ell^{m*}(\hat{\mathbf{r}}_e).\tag{F.11}$$

Albeit a laborious task, the above integral can be solved analytically (see Sect. F.1.1), giving for initial position $\mathbf{r}_0=(b,\phi_0,z_0)$ in cylindrical coordinates

$$\phi^{\pm}_{\ell m}=\frac{\mathscr{M}^{\pm}_{\ell m}}{v}K_m\left(\frac{\omega b}{v\gamma}\right)\mathrm{e}^{-im\phi_0}\mathrm{e}^{\mp i\omega z_0/v}\,.\tag{F.12}$$

Here K_m is the modified Bessel function of the second kind of Eq. (B.14), and $\gamma=1/\sqrt{1-\varepsilon\mu\beta^2}$ with $\beta=v/c$ are the Lorentz kinematic factors. Furthermore we have set

$$\mathscr{M}^{\pm}_{\ell m}=(\pm i)^{\ell+m}\sqrt{\frac{2\ell+1}{\pi}\frac{(\ell-m)!}{(\ell+m)!}\frac{(2m-1)!!}{(\sqrt{\varepsilon\mu}\gamma\beta)^m}}G_{\ell-m}^{m+1/2}\left(\frac{1}{\sqrt{\varepsilon\mu}\beta}\right),\quad\text{for } m\geq 0\tag{F.13}$$

and

$$\mathscr{M}^{\pm}_{\ell -m}=(-1)^m\mathscr{M}^{\pm}_{\ell m}\,,\tag{F.14}$$

where $G_{\ell-m}^{m+1/2}$ is the Gegenbauer polynomial. The action of $\mathbf{L}$ on $\mathbf{E}$ in Eq. (F.10) gives

$$\begin{aligned}\mathbf{L}\cdot\mathbf{E}&=\frac{e}{\varepsilon\varepsilon_0}\left\{\mathbf{L}\cdot\nabla-\frac{ik^2}{\omega}\mathbf{L}\cdot\mathbf{v}\right\}\sum_{\ell m}\phi^{+}_{\ell m}j_\ell(kr)Y_\ell^m(\hat{\mathbf{r}})\\&=-\frac{ie}{\varepsilon\varepsilon_0}\frac{vk^2}{\omega}\sum_{\ell m}\phi^{+}_{\ell m}j_\ell(kr)\mathrm{L}_zY_\ell^m(\hat{\mathbf{r}})=-\frac{ie}{\varepsilon\varepsilon_0}\frac{vk^2}{\omega}\sum_{\ell m}m\phi^{+}_{\ell m}j_\ell(kr)Y_\ell^m(\hat{\mathbf{r}})\,,\end{aligned}\tag{F.15}$$

where we used Eqs. (C.3) and (C.39a). By acting again with $\mathbf{L}$ on the general expression of the electron field (F.3), while also using Eqs. (C.26) and (C.39b), we obtain

$$\begin{aligned}\mathbf{L}\cdot\mathbf{E}&=\sum_{\ell m}\left\{b^0_{\ell m}j_\ell(kr)\mathbf{L}\cdot\mathbf{X}_{\ell m}(\hat{\mathbf{r}})+\frac{i}{k}a^0_{\ell m}\mathbf{L}\cdot\nabla\times j_\ell(kr)\mathbf{X}_{\ell m}(\hat{\mathbf{r}})\right\}\\&=\sum_{\ell m}\sqrt{\ell(\ell+1)}b^0_{\ell m}j_\ell(kr)Y_\ell^m(\hat{\mathbf{r}})\,.\end{aligned}\tag{F.16}$$

Finally, comparison of Eqs. (F.15) and (F.16) yields coefficient $b^0_{\ell m}$

$$b^0_{\ell m} = -\frac{ivk^2 e}{\varepsilon\varepsilon_0\omega}\frac{m}{\sqrt{\ell(\ell+1)}}\phi^+_{\ell m} = -\frac{ik^2 e}{\varepsilon\varepsilon_0\omega}\frac{m\mathscr{M}^+_{\ell m}}{\sqrt{\ell(\ell+1)}}K_m\left(\frac{\omega b}{v\gamma}\right)\mathrm{e}^{-im\phi_0}\mathrm{e}^{-i\omega z_0/v}\,. \tag{F.17}$$

Similarly, we find the coefficient $a^0_{\ell m}$ by acting with operator $(\mathbf{L}\times\nabla)\cdot$ on Eqs. (F.10) and (F.3). The former operation yields

$$\begin{aligned}(\mathbf{L}\times\nabla)\cdot\mathbf{E} &= \frac{e}{\varepsilon\varepsilon_0}\Big\{(\mathbf{L}\times\nabla)\cdot\nabla - \frac{ik^2}{\omega}(\mathbf{L}\times\nabla)\cdot\mathbf{v}\Big\}\sum_{\ell m}\phi^+_{\ell m}j_\ell(kr)Y^m_\ell(\hat{\mathbf{r}})\\ &= \frac{e}{\varepsilon\varepsilon_0}\frac{vk^2}{2\omega}(\mathrm{L}_+\partial_- - \mathrm{L}_-\partial_+)\sum_{\ell m}\phi^+_{\ell m}j_\ell(kr)Y^m_\ell(\hat{\mathbf{r}})\,,\end{aligned} \tag{F.18}$$

where we used Eq. (C.39b), as well as the relation

$$\mathbf{v}\cdot(\mathbf{L}\times\nabla) = \frac{iv}{2}(\mathrm{L}_+\partial_- - \mathrm{L}_-\partial_+)\,, \tag{F.19}$$

with the orbital angular momentum ladder operators $\mathrm{L}_\pm$ defined in Eq. (C.4). The right-hand side of Eq. (F.18) contains the action of the partial derivatives $\partial_\pm = \partial_x \pm i\partial_y$ on the Green's function. Notice that $\partial_\pm$ only act on x and y components, while the electron moves along the z axis, so

$$\partial_\pm\mathscr{G}[\mathbf{r}-\mathbf{r}_e(t)] = \partial_\pm\mathscr{G}[\mathbf{r}-\mathbf{r}_0-\mathbf{v}t] = -\partial^0_\pm\mathscr{G}[\mathbf{r}-\mathbf{r}_0-\mathbf{v}t]\,, \tag{F.20}$$

with $\partial^0_\pm = \partial_{x_0} \pm i\partial_{y_0}$ now involving derivatives with respect to $\mathbf{r}_0$ rather than $\mathbf{r}$ (the dependence on $\mathbf{r}_0$ is contained in $\phi^+_{\ell m}$). The action of $\partial^0_\pm$ on $\phi^+_{\ell m}$ can be derived utilizing Eq. (C.20) and the recurrence relations (B.17f), leading to

$$\partial^0_\pm\left[K_m\left(\frac{\omega b}{v\gamma}\right)\mathrm{e}^{-im\phi_0}\right] = -\frac{\omega}{v\gamma}K_{m\mp1}\left(\frac{\omega b}{v\gamma}\right)\mathrm{e}^{-i(m\mp1)\phi_0}\,, \tag{F.21a}$$

$$\partial^0_\pm\left[K_m\left(\frac{\omega b}{v\gamma}\right)\mathrm{e}^{im\phi_0}\right] = -\frac{\omega}{v\gamma}K_{m\pm1}\left(\frac{\omega b}{v\gamma}\right)\mathrm{e}^{i(m\pm1)\phi_0}\,. \tag{F.21b}$$

Substitution of Eqs. (F.20) and (F.21a) in Eq. (F.18) yields

$$\begin{aligned}(\mathbf{L}\times\nabla)\cdot\mathbf{E} &= -\frac{e}{\varepsilon\varepsilon_0}\frac{vk^2}{2\omega}\sum_{\ell m} j_\ell(kr)\big[2\alpha^m_\ell Y^{m+1}_\ell(\hat{\mathbf{r}})\partial^0_-\phi^+_{\ell m} - 2\alpha^{-m}_\ell Y^{m-1}_\ell(\hat{\mathbf{r}})\partial^0_+\phi^+_{\ell m}\big]\\ &= -\frac{e}{\varepsilon\varepsilon_0}\frac{vk^2}{2\omega}\sum_{\ell=1}^{\infty} j_\ell(kr)\Bigg[\sum_{m=-\ell+1}^{\ell+1}\big\{2\alpha^{m-1}_\ell Y^m_\ell(\hat{\mathbf{r}})\partial^0_-\phi^+_{\ell\,m-1}\big\}\end{aligned}$$

$$- \sum_{m=-\ell-1}^{\ell-1} \left\{ 2\alpha_\ell^{-m-1} Y_\ell^m(\hat{\mathbf{r}}) \partial_+^0 \phi_{\ell\, m+1}^+ \right\} \Bigg]$$

$$= -\frac{e}{\varepsilon\varepsilon_0} \frac{vk^2}{\omega} \sum_{\ell m} j_\ell(kr) Y_\ell^m(\hat{\mathbf{r}}) \left[\alpha_\ell^{-m} \partial_-^0 \phi_{\ell\, m-1}^+ - \alpha_\ell^m \partial_+^0 \phi_{\ell\, m+1}^+ \right] \tag{F.22a}$$

$$= -\frac{e}{\varepsilon\varepsilon_0} \frac{k^2}{v\gamma} \mathrm{e}^{-i\omega z_0/v} \sum_{\ell m} \mathrm{e}^{-im\phi_0} \mathscr{N}_{\ell m}^+ K_m \left(\frac{\omega b}{v\gamma} \right) j_\ell(kr) Y_\ell^m(\hat{\mathbf{r}}) \,, \tag{F.22b}$$

where we have set

$$\mathscr{N}_{\ell m}^{\pm} = \alpha_\ell^m \mathscr{M}_{\ell\, m+1}^{\pm} - \alpha_\ell^{-m} \mathscr{M}_{\ell\, m-1}^{\pm} \,. \tag{F.23}$$

To derive the last line in Eq. (F.22b) we considered $\alpha_\ell^{-\ell}$, $Y_\ell^{\pm(\ell+1)} = 0$ and $\alpha_\ell^{\pm m} = \alpha_\ell^{\mp m-1}$. By acting with operator $(\mathbf{L} \times \boldsymbol{\nabla})\cdot$ on Eq. (F.3), while also using relations (C.39b), we obtain

$$(\mathbf{L} \times \boldsymbol{\nabla}) \cdot \mathbf{E} = \sum_{\ell m} \left\{ b_{\ell m}^0 j_\ell(kr) (\mathbf{L} \times \boldsymbol{\nabla}) \cdot \mathbf{X}_{\ell m}(\hat{\mathbf{r}}) + \frac{i}{k} a_{\ell m}^0 (\mathbf{L} \times \boldsymbol{\nabla}) \cdot \boldsymbol{\nabla} \times j_\ell(kr) \mathbf{X}_{\ell m}(\hat{\mathbf{r}}) \right\}$$

$$= ik \sum_{\ell m} \sqrt{\ell(\ell+1)} a_{\ell m}^0 j_\ell(kr) Y_\ell^m(\hat{\mathbf{r}}) \,. \tag{F.24}$$

Finally, comparison of Eqs. (F.22b) and (F.24) gives

$$a_{\ell m}^0 = \frac{ik^2 e}{\varepsilon\varepsilon_0 \omega} \frac{1}{\sqrt{\varepsilon\mu}\beta\gamma} \frac{\mathscr{N}_{\ell m}^+}{\sqrt{\ell(\ell+1)}} K_m \left(\frac{\omega b}{v\gamma} \right) \mathrm{e}^{-im\phi_0} \mathrm{e}^{-i\omega z_0/v} \,. \tag{F.25}$$

F.1.1 Derivation of $\phi_{\ell m}^{\pm}$

We devote this section to deriving the analytic expression for $\phi_{\ell m}^{\pm}$ represented by the following integral

$$\phi_{\ell m}^{\pm} = ik \int_{-\infty}^{+\infty} dt\, \mathrm{e}^{\pm i\omega t} h_\ell^+(kr_e) Y_\ell^{m*}(\hat{\mathbf{r}}_e) \,.$$

For $\mathbf{r}_0 = (b, \phi_0, z_0)$ and $\mathbf{v} = v\hat{\mathbf{z}}$ in cylindrical coordinates we have

$$\phi_{\ell m}^{\pm} = ik \int_{-\infty}^{+\infty} dt\, \mathrm{e}^{\pm i\omega t} h_\ell^+ \left(k\sqrt{b^2 + [z_0 + vt]^2} \right) Y_\ell^{m*} \left(\arctan\left[\frac{z_0 + vt}{b} \right], \phi_0 \right)$$

$$= ik\mathrm{e}^{-im\phi_0} \int_{-\infty}^{+\infty} dt\, \mathrm{e}^{\pm i\omega t} h_\ell^+ \left(k\sqrt{b^2 + [z_0 + vt]^2} \right) Y_\ell^m \left(\arctan\left[\frac{z_0 + vt}{b} \right], 0 \right) . \tag{F.26}$$

The change of variable $t \rightarrow t + z_0/v$ yields

$$\phi^{\pm}_{\ell m} = ik\mathrm{e}^{-im\phi_0}\mathrm{e}^{\mp i\omega z_0/v}\int_{-\infty}^{+\infty} dt\, \mathrm{e}^{\pm i\omega t} h^{+}_{\ell}\left(k\sqrt{b^2+v^2t^2}\right) Y^m_{\ell}\left(\arctan\left[\frac{vt}{b}\right], 0\right). \tag{F.27}$$

We can identify the integrand as the term appearing in Eq. (F.6) for $\mathbf{r}' = (b, 0, vt)$, giving

$$\int_{-\infty}^{+\infty} dt\, \mathrm{e}^{\pm i\omega t}\mathscr{G}\left[\mathbf{r} - (b, 0, vt)\right]$$
$$= ik\sum_{\ell m} j_{\ell}(kr) Y^m_{\ell}(\hat{\mathbf{r}}) \int_{-\infty}^{+\infty} dt\, \mathrm{e}^{\pm i\omega t} h^{+}_{\ell}(k\sqrt{b^2+v^2t^2}) Y^m_{\ell}\left(\arctan\left[\frac{vt}{b}\right], 0\right). \tag{F.28}$$

Comparing the above equation with Eq. (F.27) we find

$$\int_{-\infty}^{+\infty} dt\, \mathrm{e}^{\pm i\omega t}\mathscr{G}\left[\mathbf{r} - (b, 0, vt)\right] = \sum_{\ell m} j_{\ell}(kr) Y^m_{\ell}(\hat{\mathbf{r}})\phi^{\pm}_{\ell m}\mathrm{e}^{im\phi_0}\mathrm{e}^{\pm i\omega z_0/v}. \tag{F.29}$$

Notice that the term on the left-hand side is the scalar potential $\varphi(\mathbf{r}, \omega)$ of Eq. (F.7a), for initial position at $\mathbf{r}_0 = (b, 0, 0)$, which we can substitute with expressions found in literature [4, 13], in particular

$$\int_{-\infty}^{+\infty} dt\, \mathrm{e}^{\pm i\omega t}\mathscr{G}\left[\mathbf{r} - (b, 0, vt)\right] = \frac{1}{2\pi v} K_0\left(\frac{\omega}{v\gamma}\sqrt{(x-b)^2+y^2}\right)\mathrm{e}^{\pm i\omega z/v}, \tag{F.30}$$

giving[6,7]

$$\frac{1}{2\pi v} K_0\left(\frac{\omega}{v\gamma}\sqrt{(x-b)^2+y^2}\right)\mathrm{e}^{\pm i\omega z/v} = \sum_{\ell m} j_{\ell}(kr) Y^m_{\ell}(\hat{\mathbf{r}})\phi^{\pm}_{\ell m}\mathrm{e}^{im\phi_0}\mathrm{e}^{\pm i\omega z_0/v}. \tag{F.31}$$

By multiplying both sides by $Y^{m'*}_{\ell'}(\hat{\mathbf{r}})$ and integrating over the solid angle one obtains

$$\phi^{\pm}_{\ell m} = \frac{1}{2\pi v}\frac{\mathrm{e}^{-im\phi_0}\mathrm{e}^{\mp i\omega z_0/v}}{j_{\ell}(kr)}\int d\Omega\, K_0\left(\frac{\omega}{v\gamma}\sqrt{(x-b)^2+y^2}\right) Y^{m*}_{\ell}(\hat{\mathbf{r}})\mathrm{e}^{\pm i\omega z/v}. \tag{F.32}$$

The integral over the azimuthal angle gives

6 Remember that Eq. (F.31) holds only provided that $v\sqrt{\varepsilon\mu} < c$ [4].

7 In principle, the angular frequency ω inside the argument of the modified Bessel function in Eq. (F.31) should read $|\omega|$ instead, to ensure the causality of the fields [13]. Since we will not be working with negative frequencies we may relax the notation.

$$\int_0^{2\pi} d\phi \, e^{-im\phi} K_0\left(\frac{\omega}{v\gamma}\sqrt{(x-b)^2+y^2}\right) = \int_0^{2\pi} d\phi \, e^{-im\phi} K_0\left(\frac{\omega}{v\gamma}\sqrt{\rho^2+b^2-2\rho b\cos\phi}\right), \tag{F.33}$$

where we set $\rho = \sqrt{x^2+y^2}$, so that $x = \rho\cos\phi$. By utilizing Eq. (B.19), the integral (F.33) takes the form

$$\int_0^{2\pi} d\phi \, e^{-im\phi} \sum_{m'=-\infty}^{\infty} K_{m'}\left(\frac{\omega b}{v\gamma}\right) I_{m'}\left(\frac{\omega\rho}{v\gamma}\right) e^{im'\phi} = 2\pi \sum_{m'=-\infty}^{\infty} I_{m'}\left(\frac{\omega\rho}{v\gamma}\right) K_{m'}\left(\frac{\omega b}{v\gamma}\right) \delta_{mm'} . \tag{F.34}$$

We then have

$$\int_0^{2\pi} d\phi \, e^{-im\phi} K_0\left(\frac{\omega}{v\gamma}\sqrt{(x-b)^2+y^2}\right) = 2\pi I_m\left(\frac{\omega\rho}{v\gamma}\right) K_m\left(\frac{\omega b}{v\gamma}\right) \tag{F.35}$$

and Eq. (F.32) reduces to

$$\phi^{\pm}_{\ell m} = \frac{(-1)^m}{v}\sqrt{\frac{2\ell+1}{4\pi}\frac{(\ell-m)!}{(\ell+m)!}} \frac{e^{-im\phi_0} e^{\mp i\omega z_0/v}}{j_\ell(kr)} K_m\left(\frac{\omega b}{v\gamma}\right) \mathscr{I}^{\pm}_{\ell m} , \tag{F.36}$$

where $\mathscr{I}^{\pm}_{\ell m}$ contains the integration over the polar angle. We transform the cylindrical coordinate system to spherical by setting $z = r\cos\theta$ and $\rho = r\sin\theta$, in which case $\mathscr{I}^{\pm}_{\ell m}$ takes the form

$$\mathscr{I}^{\pm}_{\ell m} = \int_0^{\pi} d\theta \sin\theta P^m_\ell(\cos\theta) I_m\left(\frac{\omega r\sin\theta}{v\gamma}\right) e^{\pm i\omega r\cos\theta/v} . \tag{F.37}$$

It suffices to solve the integral for $m \geq 0$, since for $m < 0$ one may use Eqs. (B.18a) and (C.8). In order to do so,[8] we introduce the Gegenbauer polynomials G^{μ}_{ν}, connected with the Legendre polynomials for $-1 < x < 1 \in \mathbb{R}$ via [2][9]

$$P^m_\ell(x) = (2m-1)!!(1-x^2)^{m/2} G^{m+1/2}_{\ell-m}(x) . \tag{F.38}$$

Using Eq. (F.38), as well as (B.13), connecting I_m and J_m, we obtain

$$\mathscr{I}^{\pm}_{\ell m} = i^{-m}(2m-1)!! \int_0^{\pi} d\theta (\sin\theta)^{m+1} G^{m+1/2}_{\ell-m}(\cos\theta) J_m\left(\frac{i\omega r\sin\theta}{v\gamma}\right) e^{\pm i\omega r\cos\theta/v} . \tag{F.39}$$

---

[8] We follow similar steps as in Ref. [14]; albeit the study pertains to the Coulomb excitation of atomic nuclei in ion collisions, the algebra is very similar.

[9] In particular we make use of relation (18.11.1) in Ref. [2] connecting the Gegenbauer polynomial with Ferrers function, which is equivalent to $(-1)^m P^m_\ell(x)$ with respect to the notation of the associated Legendre polynomials (C.7) used here.

By setting

$$\varphi = i \operatorname{arccosh}\left(\frac{1}{\sqrt{\varepsilon\mu}\beta}\right) , \tag{F.40}$$

so that $\cos\varphi = 1/(\sqrt{\varepsilon\mu}\beta)$ and $\sin\varphi = i/(\sqrt{\varepsilon\mu}\gamma\beta)$, we can modify the integral to

$$\begin{aligned} \mathscr{I}_{\ell m}^{\pm} = i^{-m}(2m-1)!! \int_0^{\pi} d\theta (\sin\theta)^{m+1} G_{\ell-m}^{m+1/2}(\cos\theta) J_m(kr\sin\theta\sin\varphi) \\ \times \left[\cos(kr\cos\theta\cos\varphi) \pm i\sin(kr\cos\theta\cos\varphi)\right] . \end{aligned} \tag{F.41}$$

Using Eqs. (7.333) found in Ref. [15] we obtain

$$\begin{aligned} \mathscr{I}_{\ell m}^{\pm} &= i^{\ell-2m}(\pm 1)^{\ell-m}(2m-1)!!(\sin\varphi)^m G_{\ell-m}^{m+1/2}(\cos\varphi)\sqrt{\frac{2\pi}{kr}} J_{\ell+1/2}(kr) \\ &= \frac{2(\pm i)^{\ell-m}(2m-1)!!}{(\sqrt{\varepsilon\mu}\gamma\beta)^m} G_{\ell-m}^{m+1/2}\left(\frac{1}{\sqrt{\varepsilon\mu}\beta}\right) j_\ell(kr) . \end{aligned} \tag{F.42}$$

Substitution of Eq. (F.42) into (F.36) yields

$$\phi_{\ell m}^{\pm} = \frac{2(\pm i)^{\ell+m}(2m-1)!!}{v(\sqrt{\varepsilon\mu}\gamma\beta)^m}\sqrt{\frac{2\ell+1}{4\pi}\frac{(\ell-m)!}{(\ell+m)!}} K_m\left(\frac{\omega b}{v\gamma}\right) G_{\ell-m}^{m+1/2}\left(\frac{1}{\sqrt{\varepsilon\mu}\beta}\right) e^{-im\phi_0} e^{\mp i\omega z_0/v} . \tag{F.43}$$

F.1.2 Bulk EELS of a Swift Electron

In the previous section we assumed an electron moving inside a non-absorbing host medium. In Sect. F.2.2, though, we consider the electron passing through the bulk of an arbitrary medium, say a dispersive and lossy metallic material, described optical parameters (ε, μ). In that case the electron beam loses energy through the excitation of bulk modes, that cannot be excited by aloof electron beams. Here, we derive the bulk EELS contribution of a swift electron.

Let us assume that the electron travels along z axis with velocity $\mathbf{v} = v\hat{\mathbf{z}}$, also passing from the axis origin, as depicted in Fig. F.1a. Then the charge density is given by $\rho_f(\mathbf{r}, t) = -e\delta(x)\delta(y)\delta(z - vt)$, or, by performing a Fourier transform with respect to x, y and t, $\rho_f(q, z, \omega) = -(e/v)e^{i\omega z/v}$. Starting from the inhomogeneous wave equation (2.23), we apply the Fourier transform with respect to x, y and t, and obtain

$$\left(-a^2 + \partial_z^2\right) \mathbf{E}(q, z, \omega) = -\frac{ie}{\varepsilon\varepsilon_0 v}\left(q\hat{\mathbf{R}} + \frac{\omega}{v\gamma^2}\hat{\mathbf{z}}\right) e^{i\omega z/v} , \tag{F.44}$$

where we set $a^2 = q^2 - k^2$ with $\mathbf{q} = (k_x, k_y, 0)$ describing momentum transverse to the electron trajectory, namely along $\hat{\mathbf{R}} = \hat{\mathbf{x}}, \hat{\mathbf{y}}$. Furthermore, k is the wavenumber and γ the Lorentz kinematic factor evaluated inside the material. Eq. (F.44) admits solutions of the form $\mathbf{E}(q, z, \omega) = \mathbf{C}_a \mathrm{e}^{i\omega z/v}$. By inserting the solution back to Eq. (F.44) we find

$$\mathbf{C}_a = \frac{ie}{\varepsilon\varepsilon_0 v} \frac{1}{[a^2 + (\omega/v)^2]} \left\{ q\hat{\mathbf{R}} + \frac{\omega}{v\gamma^2}\hat{\mathbf{z}} \right\}, \tag{F.45}$$

and

$$\mathbf{E}(q, z, \omega) = \frac{ie}{\varepsilon\varepsilon_0 v} \frac{\mathrm{e}^{i\omega z/v}}{[a^2 + (\omega/v)^2]} \left\{ q\hat{\mathbf{R}} + \frac{\omega}{v\gamma^2}\hat{\mathbf{z}} \right\}. \tag{F.46}$$

EEL Probability The electron induces polarization charges inside the medium, which in turn create electric fields that act back on it. As a result, the electron loses part of its kinetic energy due to its interaction with the environment. The energy lost is equivalent to the work $W = -e\Delta V$ (ΔV being the potential difference) corresponding to the force exerted on the electron by the induced electric field, namely

$$\Delta E = -W = e \int d\mathbf{r}_e \cdot \mathbf{E}_{\text{ind}}(\mathbf{r}_e, t) = e \int_{-\infty}^{+\infty} dt \, \mathbf{v} \cdot \mathbf{E}_{\text{ind}}(\mathbf{r}_e, t), \tag{F.47}$$

where $\mathbf{r}_e$ denotes the electron trajectory. Then, by inserting the induced electric field, Fourier-transformed with respect to t, we find

$$\Delta E = \int_0^{\infty} d\omega \hbar\omega \Gamma_{\text{EEL}}(\omega), \tag{F.48}$$

where we have defined the probability of the electron losing a quantum of energy $\hbar\omega$ as

$$\Gamma_{\text{EEL}}(\omega) = \frac{e}{\pi\hbar\omega} \int_{-\infty}^{+\infty} dt \, \mathrm{Re}\left\{ \mathrm{e}^{-i\omega t} \mathbf{v} \cdot \mathbf{E}_{\text{ind}}(\mathbf{r}_e, \omega) \right\}. \tag{F.49}$$

One may also define the momentum-resolved EEL probability $P(q, \omega)$, given by

$$P_{\text{EEL}}(q, \omega) = \frac{eq}{2\pi^2\hbar\omega} \mathrm{Re} \int dz \, \hat{\mathbf{z}} \cdot \mathbf{E}_{\text{ind}}(q, z, \omega) \mathrm{e}^{-i\omega z/v}, \tag{F.50}$$

for $\mathbf{v} = v\hat{\mathbf{z}}$, so that

$$\Gamma_{\text{EEL}}(\omega) = \int_0^{q_c} dq \, P_{\text{EEL}}(q, \omega), \tag{F.51}$$

where q_c is the momentum cutoff that prevents the above integral from going to infinity[10]. In practice, the value of q_c is related to the collection angle of the detector ϕ_{out}, as [11]

$$\hbar q_c = \sqrt{(m_e v \phi_{out})^2 + (\hbar\omega/v)^2}\,, \tag{F.52}$$

where m_e stands for the electron mass.

For an electron traveling a distance d, say from $-d/2$ to $d/2$, in the bulk of a medium, the momentum-resolved EEL probability P_{EEL} is found by inserting the electric field of Eq. (F.46) into Eq. (F.50), yielding

$$\begin{aligned} P_{EEL}(q,\omega) &= \frac{eq}{2\pi^2\hbar\omega}\mathrm{Re}\int_{-d/2}^{d/2} dz\,\frac{ie}{\varepsilon_0 v}\frac{\omega}{v}\frac{1}{[a^2+(\omega/v)^2]\gamma^2\varepsilon} \\ &= -\frac{e^2qd}{2\pi^2\hbar\varepsilon_0 v^2}\mathrm{Im}\left\{\frac{1}{[a^2+(\omega/v)^2]\gamma^2\varepsilon}\right\}. \end{aligned} \tag{F.53}$$

Then, using Eq. (F.51) we can calculate the EEL probability

$$\Gamma_{EEL}(\omega) = \int_0^{q_c} dq\, P_{EEL}(q,\omega) = -\frac{e^2 d}{2\pi^2\hbar\varepsilon_0 v^2}\mathrm{Im}\int_0^{q_c} dq\,\frac{q}{[q^2-k^2+(\omega/v)^2]\gamma^2\varepsilon}\,. \tag{F.54}$$

Using

$$\int_0^{q_c} dq\,\frac{q}{q^2+\xi} = \frac{1}{2}\Big[\ln(q^2+\xi)\Big]_0^{q_c} = \frac{1}{2}\ln\left(\frac{q_c^2}{\xi}+1\right), \tag{F.55}$$

we obtain

$$\Gamma_{EEL}(\omega) = -\frac{e^2 d}{4\pi^2\varepsilon_0\hbar v^2}\mathrm{Im}\left\{\frac{1}{\gamma^2\varepsilon}\ln\left(\left[\frac{q_c\gamma v}{\omega}\right]^2+1\right)\right\}. \tag{F.56}$$

In the case of a lossless medium $\mathrm{Im}\{\varepsilon\} = 0$, we find $\Gamma_{EEL} \neq 0$ only under the condition $v < c/\sqrt{\varepsilon\mu}$, which means that energy is lost due to generation of Cherenkov radiation.

F.2 EELS and CL of Spherical Nanoparticles

F.2.1 Aloof Electron Trajectory

In this section we provide a detailed derivation of the EEL and CL probabilities of spherical NPs excited by fast electron beams passing in aloof trajectories. The calculation is based to a large extent on Mie theory, with the excitation source being

[10] Since the entire theory is developed within LRA, it already disregards very large momentum transfers, for which the approximation fails.

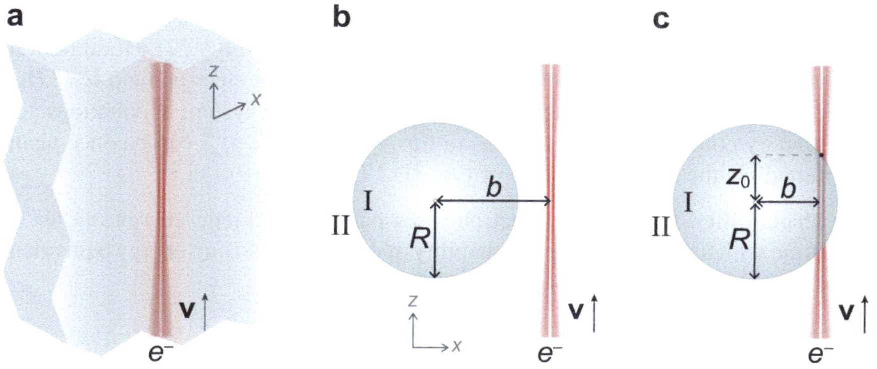

Fig. F.1 Interaction of fast electron beam with matter. Schematic illustration of a fast electron traveling with velocity $\mathbf{v} = v\hat{\mathbf{z}}$ **a** in the bulk of an unbound medium of optical parameters ε and $\mu = 1$, **b** close by a spherical NP of radius R at impact parameter $b > R$, and **c** through the NP at $b < R$, exiting at point $(x = b, z = z_0)$. In panels **b** and **c** the region inside the NP, labeled as I, is characterized by parameters ε_1 and $\mu_1 = 1$, while label II refers to the surrounding medium (air with $\varepsilon_2 = \mu_2 = 1$)

the distinguishing feature, and can still be entirely realized analytically [11, 12]. We consider the system depicted in Fig. F.1b; a spherical NP described by optical parameters (ε_1, μ_1) and embedded in a non-absorbing host medium (ε_2, μ_2) is excited by a fast electron. Following similar assumptions as in Appendix F.1, the electron trajectory is given by $\mathbf{r}_e = \mathbf{r}_0 + \mathbf{v}t$, with $r_0 = (b, \phi_0, z_0)$ in cylindrical coordinates.

Since all arguments pertaining to the spherical symmetry of the system still hold, we may follow the same steps as in Mie theory and expand the fields inside the NP $\mathbf{E}_{\text{in}}$, $\mathbf{H}_{\text{in}}$ and the surrounding medium $\mathbf{E}_{\text{out}}$, $\mathbf{H}_{\text{out}}$ on the basis of spherical waves

$$\mathbf{E}_{\text{in}}(\mathbf{r}) = \sum_{\ell m} \left\{ b^{\text{I}}_{\ell m} j_\ell(k_1 r)\mathbf{X}_{\ell m}(\hat{\mathbf{r}}) + \frac{i}{k_1} a^{\text{I}}_{\ell m} \nabla \times j_\ell(k_1 r)\mathbf{X}_{\ell m}(\hat{\mathbf{r}}) \right\}, \tag{F.57a}$$

$$\mathbf{H}_{\text{in}}(\mathbf{r}) = \frac{1}{Z_1} \sum_{\ell m} \left\{ a^{\text{I}}_{\ell m} j_\ell(k_1 r)\mathbf{X}_{\ell m}(\hat{\mathbf{r}}) - \frac{i}{k_1} b^{\text{I}}_{\ell m} \nabla \times j_\ell(k_1 r)\mathbf{X}_{\ell m}(\hat{\mathbf{r}}) \right\}, \tag{F.57b}$$

$$\mathbf{E}_{\text{inc}}(\mathbf{r}) = \sum_{\ell m} \left\{ b^{0}_{\ell m} j_\ell(k_2 r)\mathbf{X}_{\ell m}(\hat{\mathbf{r}}) + \frac{i}{k_2} a^{0}_{\ell m} \nabla \times j_\ell(k_2 r)\mathbf{X}_{\ell m}(\hat{\mathbf{r}}) \right\}, \tag{F.57c}$$

$$\mathbf{H}_{\text{inc}}(\mathbf{r}) = \frac{1}{Z_2} \sum_{\ell m} \left\{ a^{0}_{\ell m} j_\ell(k_2 r)\mathbf{X}_{\ell m}(\hat{\mathbf{r}}) - \frac{i}{k_2} b^{0}_{\ell m} \nabla \times j_\ell(k_2 r)\mathbf{X}_{\ell m}(\hat{\mathbf{r}}) \right\}, \tag{F.57d}$$

$$\mathbf{E}_{\text{scat}}(\mathbf{r}) = \sum_{\ell m} \left\{ b^{\text{II}}_{\ell m} h^+_\ell(k_2 r)\mathbf{X}_{\ell m}(\hat{\mathbf{r}}) + \frac{i}{k_2} a^{\text{II}}_{\ell m} \nabla \times h^+_\ell(k_2 r)\mathbf{X}_{\ell m}(\hat{\mathbf{r}}) \right\}, \tag{F.57e}$$

$$\mathbf{H}_{\text{scat}}(\mathbf{r}) = \frac{1}{Z_2} \sum_{\ell m} \left\{ a^{\text{II}}_{\ell m} h^+_\ell(k_2 r)\mathbf{X}_{\ell m}(\hat{\mathbf{r}}) - \frac{i}{k_2} b^{\text{II}}_{\ell m} \nabla \times h^+_\ell(k_2 r)\mathbf{X}_{\ell m}(\hat{\mathbf{r}}) \right\}. \tag{F.57f}$$

For the incident field of Eqs. (F.57c) and (F.57d), we already have derived the expansion coefficients $b^0_{\ell m}$, $a^0_{\ell m}$ in Eqs. (F.17) and (F.25), respectively[11]. The total field in region II in Fig. F.1b is equivalent to the sum of the incident electron field $\mathbf{E}_{\rm inc}$, $\mathbf{H}_{\rm inc}$ and the one scattered off the interface $\mathbf{E}_{\rm scat}$, $\mathbf{H}_{\rm scat}$. Since the system of equations and the boundary conditions is the same as in Eqs. (D.15), $b^{\rm I/II}_{\ell m}$, $a^{\rm I/II}_{\ell m}$ can be once again determined via the Mie coefficients of Eqs. (D.17) and (D.19).

EEL Probability The moving electron loses part of its kinetic energy due to its interaction with the structure. The probability of the electron losing energy equivalent to one photon $\hbar\omega$ is given by Eq. (F.49), that reads

$$\Gamma_{\rm EEL}(\omega) = \frac{e}{\pi\hbar\omega}\int_{-\infty}^{+\infty} dt\, {\rm Re}\big\{{\rm e}^{-i\omega t}\mathbf{v}\cdot\mathbf{E}_{\rm ind}(\mathbf{r}_e,\omega)\big\}\,. \tag{F.58}$$

In the case of an aloof electron trajectory the field induced by the electron to the structure in the host medium is exactly the scattered field. The above integral can be divided into its TE and TM wave contributions as $\Gamma_{\rm EEL} = \Gamma^{\rm M}_{\rm EEL} + \Gamma^{\rm E}_{\rm EEL}$. With the use of Eqs. (F.12) and (F.17) we find the contribution of the magnetic modes

$$\begin{aligned}
\Gamma^{\rm M}_{\rm EEL}(\omega) &= \frac{e}{\pi\hbar\omega}{\rm Re}\int_{-\infty}^{+\infty} dt{\rm e}^{-i\omega t}\mathbf{v}\cdot\sum_{\ell m}\frac{b^{\rm II}_{\ell m}}{\sqrt{\ell(\ell+1)}}h^+_\ell(k_2r_e)\mathbf{L}Y^m_\ell(\hat{\mathbf{r}}_e)\\
&= \frac{ve}{\pi\hbar\omega}{\rm Re}\sum_{\ell m}\frac{mT_{{\rm M}\ell}b^0_{\ell m}}{\sqrt{\ell(\ell+1)}}\int_{-\infty}^{+\infty} dt\,{\rm e}^{-i\omega t}h^+_\ell(k_2r_e)Y^m_\ell(\hat{\mathbf{r}}_e)\\
&= \frac{ve}{\pi\hbar\omega}{\rm Re}\sum_{\ell m}\frac{mT_{{\rm M}\ell}b^0_{\ell m}}{\sqrt{\ell(\ell+1)}}\frac{(-1)^m\phi^-_{\ell\,-m}}{ik_2} \qquad\qquad \text{(F.59a)}\\
&= -\frac{e^2}{c\pi\hbar\omega}\frac{\sqrt{\varepsilon_2\mu_2}}{\varepsilon_2\varepsilon_0}\sum_{\ell m}\frac{m^2}{\ell(\ell+1)}{\rm Re}\left\{T_{{\rm M}\ell}\mathscr{M}^+_{\ell m}\mathscr{M}^-_{\ell m}\left|K_m\left(\frac{\omega b}{v\gamma_2}\right)\right|^2\right\}. \qquad \text{(F.59b)}
\end{aligned}$$

Since $\mathscr{M}^-_{\ell m} = \mathscr{M}^{+\,*}_{\ell m}$, the above formula takes the form

$$\Gamma^{\rm M}_{\rm EEL} = -\frac{e^2}{\pi\varepsilon_2\varepsilon_0}\frac{\sqrt{\varepsilon_2\mu_2}}{c\hbar\omega}\sum_{\ell m}\frac{|m\mathscr{M}^+_{\ell m}|^2}{\ell(\ell+1)}\left|K_m\left(\frac{\omega b}{v\gamma_2}\right)\right|^2{\rm Re}\{T_{{\rm M}\ell}\}\,. \tag{F.60}$$

The contribution of the electric modes is equal to

$$\Gamma^{\rm E}_{\rm EEL}(\omega) = \frac{e}{\pi\hbar\omega}{\rm Re}\int_{-\infty}^{+\infty} dt\,{\rm e}^{-i\omega t}\mathbf{v}\cdot\sum_{\ell m}\frac{i}{k_2}\frac{a^{\rm II}_{\ell m}}{\sqrt{\ell(\ell+1)}}(\nabla\times\mathbf{L})h^+_\ell(k_2r)Y^m_\ell(\hat{\mathbf{r}})\Big|_{\mathbf{r}=\mathbf{r}_e}$$

---

[11] We note here that Eqs. (F.57c) and (F.57d) hold only at $r < r_e$, which is sufficient for calculating the EEL and CL probability.

$$= \frac{e}{\pi\hbar\omega}\mathrm{Re}\sum_{\ell m}\int_{-\infty}^{+\infty} dt\, e^{-i\omega t}\frac{i}{k_2}\frac{a^{\mathrm{II}}_{\ell m}}{\sqrt{\ell(\ell+1)}}\frac{-iv}{2}(\partial_- \mathrm{L}_+ - \partial_+ \mathrm{L}_-)h^+_\ell(k_2 r)Y^m_\ell(\hat{\mathbf{r}})\Big|_{\mathbf{r}=\mathbf{r}_e}$$

$$= \frac{e}{\pi\hbar\omega}\frac{v}{k_2}\mathrm{Re}\sum_{\ell m}\frac{a^{\mathrm{II}}_{\ell m}}{\sqrt{\ell(\ell+1)}}$$

$$\times\int_{-\infty}^{+\infty} dt\, e^{-i\omega t}\big[\alpha^m_\ell \partial_- Y^{m+1}_\ell(\hat{\mathbf{r}}) - \alpha^{-m}_\ell \partial_+ Y^{m-1}_\ell(\hat{\mathbf{r}})\big]h^+_\ell(k_2 r)\Big|_{\mathbf{r}=\mathbf{r}_e}, \tag{F.61}$$

where the partial derivatives $\partial_\pm = \partial_x \pm i\partial_y$ act on functions evaluated at $\mathbf{r}_e = \mathbf{r}_0 + \mathbf{v}t$. Since the vectors $\mathbf{r}_e$ and $\mathbf{r}_0$ only differ in their z component, we may replace $\partial_\pm$ by $\partial^0_\pm$, defined as in Eq. (C.20). Using Eqs. (F.12) and (F.25), we obtain

$$\Gamma^{\mathrm{E}}_{\mathrm{EEL}}(\omega) = \frac{e}{\pi\hbar\omega}\frac{v}{k_2}\mathrm{Re}\sum_{\ell m}\frac{T_{\mathrm{E}\ell}a^0_{\ell m}}{\sqrt{\ell(\ell+1)}}\left[\alpha^m_\ell\partial^0_-\frac{(-1)^{m+1}\phi^-_{\ell-(m+1)}}{ik_2} - \alpha^{-m}_\ell\partial^0_+\frac{(-1)^{m-1}\phi^-_{\ell-(m-1)}}{ik_2}\right]$$

$$= -\frac{e^2}{c\pi\hbar\omega}\frac{1}{\varepsilon_2\varepsilon_0}\frac{1}{\sqrt{\varepsilon_2\mu_2\beta^2\gamma_2^2}}\sum_{\ell m}\frac{1}{\ell(\ell+1)}\mathrm{Re}\left\{T_{\mathrm{E}\ell}\mathscr{N}^-_{\ell m}\mathscr{N}^+_{\ell m}\left|K_m\left(\frac{\omega b}{v\gamma_2}\right)\right|^2\right\}$$

$$= -\frac{e^2}{\pi\varepsilon_2\varepsilon_0}\frac{1}{c\hbar\omega}\frac{1}{\sqrt{\varepsilon_2\mu_2\beta^2\gamma_2^2}}\sum_{\ell m}\frac{|\mathscr{N}^+_{\ell m}|^2}{\ell(\ell+1)}\left|K_m\left(\frac{\omega b}{v\gamma_2}\right)\right|^2\mathrm{Re}\{T_{\mathrm{E}\ell}\}, \tag{F.62}$$

where we invoked that $\mathscr{N}^-_{\ell m} = \mathscr{N}^{+*}_{\ell m}$. Finally, Eqs. (F.60) and (F.62) give the total contribution

$$\Gamma_{\mathrm{EEL}}(\omega) = \frac{e^2}{\pi\varepsilon_2\varepsilon_0}\frac{\sqrt{\varepsilon_2\mu_2}}{c\hbar\omega}\sum_{\ell m}\frac{|K_m(\omega b/[v\gamma_2])|^2}{\ell(\ell+1)}\Bigg[\left|m\mathscr{M}^+_{\ell m}\right|^2\mathrm{Re}\{T_{\mathrm{M}\ell}\}$$

$$+\left|\frac{\mathscr{N}^+_{\ell m}}{\sqrt{\varepsilon_2\mu_2}\beta\gamma_2}\right|^2\mathrm{Re}\{T_{\mathrm{E}\ell}\}\Bigg]. \tag{F.63}$$

CL Probability Part of the electron kinetic energy that is transferred to the structure is radiated back to the environment. The radiated energy can be calculated by integrating the Poynting vector normal to the surface of an arbitrarily large sphere ($kr \to \infty$) centered around the NP

$$\Delta E_{\mathrm{CL}} = \int_0^\infty dt \int d\Omega\, r^2[\mathbf{E}_{\mathrm{ind}}(\mathbf{r},t)\times\mathbf{H}_{\mathrm{ind}}(\mathbf{r},t)]\cdot\hat{\mathbf{r}}\,. \tag{F.64}$$

By substituting the Fourier-transformed field we obtain

$$\Delta E_{\mathrm{CL}} = \frac{1}{\pi}\int_0^\infty d\omega\hbar\omega\Gamma_{\mathrm{CL}}(\omega)\,, \tag{F.65}$$

where Γ_{CL} expresses the probability of collecting a photon of energy $\hbar\omega$ in the far field, defined as

$$\Gamma_{\rm CL}(\omega) = \frac{r^2}{\pi\hbar\omega}\int d\Omega\,{\rm Re}\big\{\mathbf{E}_{\rm ind}(\mathbf{r},\omega)\times\mathbf{H}^*_{\rm ind}(\mathbf{r},\omega)\big\}\cdot\hat{\mathbf{r}}\,. \tag{F.66}$$

The induced field that radiates to the far field is equivalent to the scattered field. We therefore insert the scattered fields of Eqs. (F.57e) and (F.57f) in Eq. (F.66), and following similar algebra as we did for the derivation of Eq. (D.21a), we find

$$\Gamma_{\rm CL}(\omega) = \frac{1}{\pi\hbar\omega Z_2 k_2^2}\sum_{\ell m}\Big[\big|T_{{\rm M}\ell}\big|^2\big|b^0_{\ell m}\big|^2 + \big|T_{{\rm E}\ell}\big|^2\big|a^0_{\ell m}\big|^2\Big]\,. \tag{F.67}$$

As a final step we insert the expansion coefficients of Eqs. (F.25) and (F.17) back to Eq. (F.67) and obtain

$$\Gamma_{\rm CL}(\omega) = \frac{e^2}{\pi\varepsilon_2\varepsilon_0}\frac{\sqrt{\varepsilon_2\mu_2}}{c\hbar\omega}\sum_{\ell m}\frac{|K_m(\omega b/[v\gamma_2])|^2}{\ell(\ell+1)}\left[\big|m\,\mathscr{M}^+_{\ell m}\big|^2\big|T_{{\rm M}\ell}\big|^2 + \left|\frac{\mathscr{N}^+_{\ell m}}{\sqrt{\varepsilon_2\mu_2}\beta\gamma_2}\right|^2\big|T_{{\rm E}\ell}\big|^2\right]\,. \tag{F.68}$$

F.2.2 Penetrating Electron Trajectory

Let us consider now a spherical NP excited by an electron beam that passes through the bulk of the material, entering at point $(b, \phi_0, -z_0)$ and exiting at (b, ϕ_0, z_0) in cylindrical coordinates[12] as illustrated in Fig. F.1c. In order to solve the boundary problem we need to specify the fields very close to the NP boundary. The incident field originating from the part of the electron trajectory lying externally to the NP is once again given by Eq. (F.10)

$$\mathbf{E}^{\rm II}_e(\mathbf{r},\omega) = \frac{e}{\varepsilon_2\varepsilon_0}\Big\{\nabla - \frac{ik_2^2}{\omega}\mathbf{v}\Big\}\sum_{\ell m}\phi^{\rm II}_{\ell m}\,j_\ell(k_2 r)Y^m_\ell(\hat{\mathbf{r}})\,, \tag{F.69}$$

but, now

$$\phi^{\rm II}_{\ell m} = ik_2\int_{-\infty}^{+\infty} dt\,e^{i\omega t}h^+_\ell(k_2 r_e)Y^{m*}_\ell(\hat{\mathbf{r}}_e) - ik_2\int_{t_1}^{t_2} dt\,e^{i\omega t}h^+_\ell(k_2 r_e)Y^{m*}_\ell(\hat{\mathbf{r}}_e)\,, \tag{F.70}$$

where $t_{1/2}$ is the time at which the electron enters/exits the NP. The first term of the expression above is exactly Eq. (F.11). In the second term changing the integration variable to $z = vt$ yields

[12] Notice the change of notation; parameter z_0 refers now to the z coordinate of the electron exit point on the surface of the NP, rather than the starting point of the electron trajectory. Regarding the latter, we consider that the electron travels from $z\to-\infty$ to ∞.

$$\int_{t_1}^{t_2} dt\, \mathrm{e}^{\pm i\omega t} h_\ell^+(k_2 r_e) Y_\ell^{m*}(\hat{\mathbf{r}}_e) = \frac{\mathrm{e}^{-im\phi_0}}{v} \int_{-z_0}^{z_0} dz\, \mathrm{e}^{i\omega z/v} h_\ell^+(k_2 r) Y_\ell^m(\theta, 0) \ , \quad \text{(F.71)}$$

where $z_0 = \sqrt{R^2 - b^2}$, $r = \sqrt{b^2 + z^2}$ and $\theta = \arccos(z/r)$. Therefore, $\phi_{\ell m}^{\rm II}$ takes the form

$$\phi_{\ell m}^{\rm II} = \frac{\mathrm{e}^{-im\phi_0}}{v} \left[\mathscr{M}_{\ell m}^+ K_m \left(\frac{\omega b}{v\gamma_2} \right) - ik_2 \int_{-z_0}^{z_0} dz\, \mathrm{e}^{i\omega z/v} h_\ell^+(k_2 r) Y_\ell^m(\theta, 0) \right] . \quad \text{(F.72)}$$

The $b_{\ell m}^{0,\rm II}$ coefficient can be easily derived by inserting $\phi_{\ell m}^{\rm II}$ into Eq. (F.17), giving

$$b_{\ell m}^{0,\rm II} = -\frac{ik_2^2 e}{\varepsilon_2 \varepsilon_0 \omega} \frac{m\mathrm{e}^{-im\phi_0}}{\sqrt{\ell(\ell+1)}} \left[\mathscr{M}_{\ell m}^+ K_m \left(\frac{\omega b}{v\gamma_2} \right) - ik_2 \int_{-z_0}^{z_0} dz\, \mathrm{e}^{i\omega z/v} h_\ell^+(k_2 r) Y_\ell^m(\theta, 0) \right] . \quad \text{(F.73)}$$

To derive coefficient $a_{\ell m}^{0,\rm II}$ we follow the same basic steps as in Appendix F.2.1. Recalling Eq. (F.22a), we have

$$(\mathbf{L} \times \nabla) \cdot \mathbf{E}_e^{\rm II} = -\frac{e}{\varepsilon_2 \varepsilon_0} \frac{v k_2^2}{\omega} \sum_{\ell m} j_\ell(kr) Y_\ell^m(\hat{\mathbf{r}}) \left[\alpha_\ell^{-m} \partial_-^0 \phi_{\ell\, m-1}^{\rm II} - \alpha_\ell^m \partial_+^0 \phi_{\ell\, m+1}^{\rm II} \right] . \quad \text{(F.74)}$$

The action of the partial derivative operators on the first term of $\phi_{\ell m}^{\rm II}$ yields Eq. (F.22b), and using Eq. (C.21) we find

$$\begin{aligned}(\mathbf{L} \times \nabla) \cdot \mathbf{E}_e^{\rm II} = -\frac{e}{\varepsilon_2 \varepsilon_0} \frac{k_2^2}{\omega} \sum_{\ell m} \mathrm{e}^{-im\phi_0} j_\ell(k_2 r) Y_\ell^m(\hat{\mathbf{r}}) \Bigg[\frac{\omega}{v\gamma_2} \mathscr{N}_{\ell m}^+ K_m \left(\frac{\omega b}{v\gamma_2} \right) \\ - \frac{ik_2}{b} \int_{-z_0}^{z_0} dz\, \mathrm{e}^{i\omega z/v} \left\{ \mathscr{H}_{\ell m}^+(k_2 z) + \mathscr{H}_{\ell m}^-(k_2 z) \right\} \Bigg] , \quad \text{(F.75)}\end{aligned}$$

where we have set

$$\begin{aligned}\mathscr{F}_{\ell m}^{\pm}(k_n z) = \mp \alpha_\ell^{\pm m} \Big\{ (1 \pm m) f_\ell(k_n r) Y_\ell^{m\pm 1}(\theta, 0) + \frac{k_n b^2}{r} f_\ell'(k_n r) Y_\ell^{m\pm 1}(\theta, 0) \\ \pm \frac{zb}{r^2} f_\ell(k_n r) \left[\alpha_\ell^{\pm m+1} Y_\ell^{m\pm 2}(\theta, 0) - \alpha_\ell^{\pm m} Y_\ell^m(\theta, 0) \right] \Big\} , \quad \text{(F.76)}\end{aligned}$$

with $\mathscr{F} = \mathscr{H}, \mathscr{J}$ corresponding to formulas containing $f_\ell = h_\ell^+, j_\ell$, while the index $n = 1, 2$ refers to the material. Comparison of Eqs. (F.24) and (F.75) yields

$$\begin{aligned}a_{\ell m}^{0,\rm II} = \frac{ik_2^2 e}{\varepsilon_2 \varepsilon_0 \omega} \frac{\mathrm{e}^{-im\phi_0}}{\sqrt{\ell(\ell+1)}} \Bigg[\frac{\mathscr{N}_{\ell m}^+}{\sqrt{\varepsilon_2 \mu_2} \beta \gamma_2} K_m \left(\frac{\omega b}{v\gamma_2} \right) \\ - \frac{i}{b} \int_{-z_0}^{z_0} dz\, \mathrm{e}^{i\omega z/v} \left\{ \mathscr{H}_{\ell m}^+(k_2 z) + \mathscr{H}_{\ell m}^-(k_2 z) \right\} \Bigg] . \quad \text{(F.77)}\end{aligned}$$

According to Eqs. (F.6) and (F.9), the electric field corresponding to the electron trajectory that lies inside the NP, i.e. for $r_e < R$, takes the form

$$\mathbf{E}_e^{\mathrm{I}}(\mathbf{r},\omega) = \frac{e}{\varepsilon_1\varepsilon_0}\left\{\nabla - \frac{ik_1^2}{\omega}\mathbf{v}\right\}\sum_{\ell m}\phi_{\ell m}^{\mathrm{I}} h_\ell^+(k_1 r) Y_\ell^m(\hat{\mathbf{r}}) , \tag{F.78}$$

with

$$\phi_{\ell m}^{\mathrm{I}} = ik_1\int_{t_1}^{t_2} dt\, \mathrm{e}^{i\omega t} j_\ell(k_1 r_e) Y_\ell^{m*}(\hat{\mathbf{r}}_e) = \frac{ik_1}{v}\mathrm{e}^{-im\phi_0}\int_{-z_0}^{z_0} dz\, \mathrm{e}^{i\omega z/v} j_\ell(k_1 r) Y_\ell^m(\theta, 0) . \tag{F.79}$$

It is then straightforward that the TE and TM expansion coefficients are equivalent to

$$b_{\ell m}^{0,\mathrm{I}} = -\frac{ik_1^2 e}{\varepsilon_1\varepsilon_0\omega}\frac{m\mathrm{e}^{-im\phi_0}}{\sqrt{\ell(\ell+1)}} ik_1 \int_{-z_0}^{z_0} dz\, \mathrm{e}^{i\omega z/v} j_\ell(k_1 r) Y_\ell^m(\theta, 0) , \tag{F.80}$$

and

$$a_{\ell m}^{0,\mathrm{I}} = \frac{ik_1^2 e}{\varepsilon_1\varepsilon_0\omega}\frac{\mathrm{e}^{-im\phi_0}}{\sqrt{\ell(\ell+1)}}\frac{i}{b}\int_{-z_0}^{z_0} dz\, \mathrm{e}^{i\omega z/v}\left\{\mathscr{I}_{\ell m}^-(k_1 z) + \mathscr{I}_{\ell m}^+(k_1 z)\right\} . \tag{F.81}$$

Let us now write the general expressions of the total fields that hold in regions I and II, very close to the boundary. In each region the fields will be the sum of the electron field $\mathbf{E}_e^{\mathrm{I/II}}$, $\mathbf{H}_e^{\mathrm{I/II}}$ and the fields induced due to the presence of a boundary $\mathbf{E}_{\mathrm{B}}^{\mathrm{I/II}}$, $\mathbf{H}_{\mathrm{B}}^{\mathrm{I/II}}$,

$$\mathbf{E}_{\mathrm{tot}}^{\mathrm{I}}(\mathbf{r}) = \sum_{\ell m}\Big\{b_{\ell m}^{0,\mathrm{I}} h_\ell^+(k_1 r)\mathbf{X}_{\ell m}(\hat{\mathbf{r}}) + \frac{i}{k_1} a_{\ell m}^{0,\mathrm{I}}\nabla\times h_\ell^+(k_1 r)\mathbf{X}_{\ell m}(\hat{\mathbf{r}})$$
$$+ b_{\ell m}^{\mathrm{I}} j_\ell(k_1 r)\mathbf{X}_{\ell m}(\hat{\mathbf{r}}) + \frac{i}{k_1} a_{\ell m}^{\mathrm{I}}\nabla\times j_\ell(k_1 r)\mathbf{X}_{\ell m}(\hat{\mathbf{r}})\Big\} , \tag{F.82a}$$

$$\mathbf{H}_{\mathrm{tot}}^{\mathrm{I}}(\mathbf{r}) = \frac{1}{Z_1}\sum_{\ell m}\Big\{a_{\ell m}^{0,\mathrm{I}} j_\ell(k_1 r)\mathbf{X}_{\ell m}(\hat{\mathbf{r}}) - \frac{i}{k_1} b_{\ell m}^{0,\mathrm{I}}\nabla\times j_\ell(k_1 r)\mathbf{X}_{\ell m}(\hat{\mathbf{r}})$$
$$+ a_{\ell m}^{\mathrm{I}} j_\ell(k_1 r)\mathbf{X}_{\ell m}(\hat{\mathbf{r}}) - \frac{i}{k_1} b_{\ell m}^{\mathrm{I}}\nabla\times j_\ell(k_1 r)\mathbf{X}_{\ell m}(\hat{\mathbf{r}})\Big\} , \tag{F.82b}$$

$$\mathbf{E}_{\mathrm{tot}}^{\mathrm{II}}(\mathbf{r}) = \sum_{\ell m}\Big\{b_{\ell m}^{0,\mathrm{II}} j_\ell(k_2 r)\mathbf{X}_{\ell m}(\hat{\mathbf{r}}) + \frac{i}{k_2} a_{\ell m}^{0,\mathrm{II}}\nabla\times j_\ell(k_2 r)\mathbf{X}_{\ell m}(\hat{\mathbf{r}})$$
$$+ b_{\ell m}^{\mathrm{II}} h_\ell^+(k_2 r)\mathbf{X}_{\ell m}(\hat{\mathbf{r}}) + \frac{i}{k_2} a_{\ell m}^{\mathrm{II}}\nabla\times h_\ell^+(k_2 r)\mathbf{X}_{\ell m}(\hat{\mathbf{r}})\Big\} , \tag{F.82c}$$

$$\mathbf{H}^{\mathrm{II}}_{\mathrm{tot}}(\mathbf{r}) = \frac{1}{Z_2}\sum_{\ell m}\Big\{a^{0,\mathrm{II}}_{\ell m} j_\ell(k_2 r)\mathbf{X}_{\ell m}(\hat{\mathbf{r}}) - \frac{i}{k_2} b^{0,\mathrm{II}}_{\ell m}\nabla\times j_\ell(k_2 r)\mathbf{X}_{\ell m}(\hat{\mathbf{r}})$$
$$+ a^{\mathrm{II}}_{\ell m} h^+_\ell(k_2 r)\mathbf{X}_{\ell m}(\hat{\mathbf{r}}) - \frac{i}{k_2} b^{\mathrm{II}}_{\ell m}\nabla\times h^+_\ell(k_2 r)\mathbf{X}_{\ell m}(\hat{\mathbf{r}})\Big\}\,. \qquad \text{(F.82d)}$$

To determine the coefficients $a^{\mathrm{I/II}}_{\ell m}$ and $b^{\mathrm{I/II}}_{\ell m}$ we invoke the boundary conditions (D.15). Utilizing relations (C.31) and (C.40), we find

$$b^{0,\mathrm{I}}_{\ell m} h^+_\ell(k_1 R) + b^{\mathrm{I}}_{\ell m} j_\ell(k_1 R) = b^{0,\mathrm{II}}_{\ell m} j_\ell(k_2 R) + b^{\mathrm{II}}_{\ell m} h^+_\ell(k_2 R)\,, \qquad \text{(F.83a)}$$

$$\frac{1}{Z_1} a^{0,\mathrm{I}}_{\ell m} h^+_\ell(k_1 R) + \frac{1}{Z_1} a^{\mathrm{I}}_{\ell m} j_\ell(k_1 R) = \frac{1}{Z_2} a^{0,\mathrm{II}}_{\ell m} j_\ell(k_2 R) + \frac{1}{Z_2} a^{\mathrm{II}}_{\ell m} h^+_\ell(k_2 R)\,, \qquad \text{(F.83b)}$$

$$\frac{1}{k_1} a^{0,\mathrm{I}}_{\ell m} \xi'_\ell(k_1 R) + \frac{1}{k_1} a^{\mathrm{I}}_{\ell m} \Psi'_\ell(k_1 R) = \frac{1}{k_2} a^{0,\mathrm{II}}_{\ell m} \Psi'_\ell(k_2 R) + \frac{1}{k_2} a^{\mathrm{II}}_{\ell m} \xi'_\ell(k_2 R)\,, \qquad \text{(F.83c)}$$

$$\frac{1}{Z_1 k_1} b^{0,\mathrm{I}}_{\ell m} \xi'_\ell(k_1 R) + \frac{1}{Z_1 k_1} b^{\mathrm{I}}_{\ell m} \Psi'_\ell(k_1 R) = \frac{1}{Z_2 k_2} b^{0,\mathrm{II}}_{\ell m} \Psi'_\ell(k_2 R) + \frac{1}{Z_2 k_2} b^{\mathrm{II}}_{\ell m} \xi'_\ell(k_2 R)\,. \qquad \text{(F.83d)}$$

The above system of equations leads to

$$a^{\mathrm{I}}_{\ell m} = T^{11}_{\mathrm{E}\ell} a^{0,\mathrm{I}}_{\ell m} + T^{21}_{\mathrm{E}\ell} a^{0,\mathrm{II}}_{\ell m}\,, \qquad \text{(F.84a)}$$
$$a^{\mathrm{II}}_{\ell m} = T^{12}_{\mathrm{E}\ell} a^{0,\mathrm{I}}_{\ell m} + T^{22}_{\mathrm{E}\ell} a^{0,\mathrm{II}}_{\ell m}\,, \qquad \text{(F.84b)}$$
$$b^{\mathrm{I}}_{\ell m} = T^{11}_{\mathrm{M}\ell} b^{0,\mathrm{I}}_{\ell m} + T^{21}_{\mathrm{M}\ell} b^{0,\mathrm{II}}_{\ell m}\,, \qquad \text{(F.84c)}$$
$$b^{\mathrm{II}}_{\ell m} = T^{12}_{\mathrm{M}\ell} b^{0,\mathrm{I}}_{\ell m} + T^{22}_{\mathrm{M}\ell} b^{0,\mathrm{II}}_{\ell m}\,, \qquad \text{(F.84d)}$$

where $T^{22}_{\mathrm{E/M}\ell}$,[13] $T^{21}_{\mathrm{E/M}\ell}$ are given by Eqs. (D.17), (D.18), (D.19), while $T^{11}_{\mathrm{E/M}\ell}$, $T^{12}_{\mathrm{E/M}\ell}$ read

$$T^{11}_{\mathrm{E}\ell} = \frac{a^{\mathrm{I}}_{\ell m}}{a^{0,\mathrm{I}}_{\ell m}} = \frac{\varepsilon_2 h^+_\ell(k_2 R)\xi'_\ell(k_1 R) - \varepsilon_1 \xi'_\ell(k_2 R) h^+_\ell(k_1 R)}{\varepsilon_1 \xi'_\ell(k_2 R) j_\ell(k_1 R) - \varepsilon_2 h^+_\ell(k_2 R)\Psi'_\ell(k_1 R)}\,, \qquad \text{(F.85a)}$$

$$T^{11}_{\mathrm{M}\ell} = \frac{b^{\mathrm{I}}_{\ell m}}{b^{0,\mathrm{I}}_{\ell m}} = \frac{\mu_2 h^+_\ell(k_2 R)\xi'_\ell(k_1 R) - \mu_1 \xi'_\ell(k_2 R) h^+_\ell(k_1 R)}{\mu_1 \xi'_\ell(k_2 R) j_\ell(k_1 R) - \mu_2 h^+_\ell(k_2 R)\Psi'_\ell(k_1 R)}\,, \qquad \text{(F.85b)}$$

$$T^{12}_{\mathrm{E}\ell} = \frac{a^{\mathrm{II}}_{\ell m}}{a^{0,\mathrm{I}}_{\ell m}} = \frac{\varepsilon_1 k_2}{k_1}\,\frac{h^+_\ell(k_1 R)\Psi'_\ell(k_1 R) - \xi'_\ell(k_1 R) j_\ell(k_1 R)}{\varepsilon_1 \xi'_\ell(k_2 R) j_\ell(k_1 R) - \varepsilon_2 h^+_\ell(k_2 R)\Psi'_\ell(k_1 R)}\,, \qquad \text{(F.86a)}$$

[13] Notice that we changed the notation regarding the Mie coefficients of the scattered field of Eqs. (D.17) from $T_{\mathrm{E/M}\ell}$ to $T^{22}_{\mathrm{E/M}\ell}$.

$$T_{\mathrm{M}\ell}^{12} = \frac{b_{\ell m}^{\mathrm{II}}}{b_{\ell m}^{0,\mathrm{I}}} = \mu_2 \frac{h_\ell^+(k_1 R)\Psi_\ell'(k_1 R) - \xi_\ell'(k_1 R) j_\ell(k_1 R)}{\mu_1 \xi_\ell'(k_2 R) j_\ell(k_1 R) - \mu_2 h_\ell^+(k_2 R)\Psi_\ell'(k_1 R)} . \tag{F.86b}$$

Using the Wronskian determinant of Eq. (B.10a), the last two equations can be written as

$$T_{\mathrm{E}\ell}^{12} = \frac{i}{k_1 R}\sqrt{\frac{\varepsilon_2\varepsilon_1\mu_2}{\mu_1}} \frac{1}{\varepsilon_1 \xi_\ell'(k_2 R) j_\ell(k_1 R) - \varepsilon_2 h_\ell^+(k_2 R)\Psi_\ell'(k_1 R)} , \tag{F.87a}$$

$$T_{\mathrm{M}\ell}^{12} = \frac{i\mu_2}{k_1 R} \frac{1}{\mu_1 \xi_\ell'(k_2 R) j_\ell(k_1 R) - \mu_2 h_\ell^+(k_2 R)\Psi_\ell'(k_1 R)} . \tag{F.87b}$$

EEL Probability The probability of the electron losing a quantum of energy $\hbar\omega$ due to the interaction is given by Eq. (F.49)

$$\begin{aligned}\Gamma_{\mathrm{EEL}}(\omega) &= \frac{e}{\pi\hbar\omega}\mathrm{Re}\int_{-\infty}^{+\infty} dt\, \mathrm{e}^{-i\omega t}\mathbf{v}\cdot\mathbf{E}_{\mathrm{ind}}(\mathbf{r}_e,\omega) = \frac{e}{\pi\hbar\omega}\mathrm{Re}\Bigg[\int_{-\infty}^{t_1} dt\, \mathrm{e}^{-i\omega t}\mathbf{v}\cdot\mathbf{E}_{\mathrm{ind}}^{\mathrm{II}}(\mathbf{r}_e,\omega) \\ &\quad + \int_{t_1}^{t_2} dt\, \mathrm{e}^{-i\omega t}\mathbf{v}\cdot\mathbf{E}_{\mathrm{ind}}^{\mathrm{I}}(\mathbf{r}_e,\omega) + \int_{t_2}^{\infty} dt\,\left\{\mathrm{e}^{-i\omega t}\mathbf{v}\cdot\mathbf{E}_{\mathrm{ind}}^{\mathrm{II}}(\mathbf{r}_e,\omega)\right\}\Bigg] \\ &= \frac{e}{\pi\hbar\omega}\mathrm{Re}\Bigg[\int_{-\infty}^{+\infty} dt\, \mathrm{e}^{-i\omega t}\mathbf{v}\cdot\mathbf{E}_{\mathrm{ind}}^{\mathrm{II}}(\mathbf{r}_e,\omega) - \int_{t_1}^{t_2} dt\, \mathrm{e}^{-i\omega t}\mathbf{v}\cdot\mathbf{E}_{\mathrm{ind}}^{\mathrm{II}}(\mathbf{r}_e,\omega) \\ &\quad + \int_{t_1}^{t_2} dt\, \mathrm{e}^{-i\omega t}\mathbf{v}\cdot\mathbf{E}_{\mathrm{ind}}^{\mathrm{I}}(\mathbf{r}_e,\omega)\Bigg] ,\end{aligned} \tag{F.88}$$

where the induced field in this case is defined as the total field minus the electron field $\mathbf{E}_0$ in the absence of the structure, i.e. $\mathbf{E}_{\mathrm{ind}}^{\mathrm{I/II}} = \mathbf{E}_{\mathrm{tot}}^{\mathrm{I/II}} - \mathbf{E}_0$. Since $\mathbf{E}_e^{\mathrm{II}} = \mathbf{E}_0$, the induced field in region II is equivalent to the scattered field. We may split the contribution to the EEL probability into the one caused by the external part of the electron trajectory, and the one that lies inside the structure $\Gamma_{\mathrm{EEL}} = \Gamma_{\mathrm{II}} + \Gamma_{\mathrm{I}}$, so that

$$\begin{aligned}\Gamma_{\mathrm{EEL}}(\omega) &= \Gamma_{\mathrm{II}}(\omega) + \Gamma_{\mathrm{I}}(\omega) \\ &= \frac{e}{\pi\hbar\omega}\mathrm{Re}\Bigg[\int_{-\infty}^{+\infty} dt\, \mathrm{e}^{-i\omega t}\mathbf{v}\cdot\mathbf{E}_{\mathrm{B}}^{\mathrm{II}}(\mathbf{r}_e,\omega) - \int_{t_1}^{t_2} dt\, \mathrm{e}^{-i\omega t}\mathbf{v}\cdot\mathbf{E}_{\mathrm{B}}^{\mathrm{II}}(\mathbf{r}_e,\omega) \\ &\quad + \int_{t_1}^{t_2} dt\, \mathrm{e}^{-i\omega t}\mathbf{v}\cdot\left\{\mathbf{E}_e^{\mathrm{I}}(\mathbf{r}_e,\omega) + \mathbf{E}_{\mathrm{B}}^{\mathrm{I}}(\mathbf{r}_e,\omega) - \mathbf{E}_0(\mathbf{r}_e,\omega)\right\}\Bigg] .\end{aligned} \tag{F.89}$$

We first calculate the contribution of the magnetic modes to Γ_{II}. Using Eqs. (F.59a) and (F.72), we obtain

$$\Gamma_{\mathrm{II}}^{\mathrm{M}}(\omega) = \frac{ve}{\pi\hbar\omega}\mathrm{Re}\sum_{\ell m} \frac{m b_{\ell m}^{\mathrm{II}}}{\sqrt{\ell(\ell+1)}} \frac{(-1)^m \phi_{\ell -m}^{\mathrm{II}-}}{ik_2}$$

$$= \frac{e}{\pi\hbar\omega}\text{Re}\sum_{\ell m} e^{im\phi_0}\frac{mb^{\text{II}}_{\ell m}}{\sqrt{\ell(\ell+1)}}\left[\frac{\mathscr{M}^{+\,*}_{\ell m}}{ik_2}K_m\left(\frac{\omega b}{v\gamma_2}\right) - \int_{-z_0}^{z_0} dz\, e^{-i\omega z/v} h^+_\ell(k_2 r) Y^m_\ell(\theta, 0)\right]. \tag{F.90}$$

Next, we calculate the contribution of the electric modes, following the same algebraic recipe as for the derivation of Eq. (F.62), and find

$$\begin{aligned}
\Gamma^{\text{E}}_{\text{II}}(\omega) &= \frac{e}{\pi\hbar\omega}\frac{v}{k_2}\text{Re}\sum_{\ell m}\frac{a^{\text{II}}_{\ell m}}{\sqrt{\ell(\ell+1)}} \\
&\quad \left[\alpha^m_\ell \partial^0_- \frac{(-1)^{m+1}\phi^{\text{II}-}_{\ell\,-(m+1)}}{ik_2} - \alpha^{-m}_\ell \partial^0_+ \frac{(-1)^{m-1}\phi^{\text{II}-}_{\ell\,-(m-1)}}{ik_2}\right] \\
&= -\frac{e}{\pi\hbar\omega}\text{Re}\sum_{\ell m} e^{im\phi_0}\frac{a^{\text{II}}_{\ell m}}{\sqrt{\ell(\ell+1)}} \\
&\quad \left[\frac{\mathscr{N}^{+\,*}_{\ell m}}{ik_2\sqrt{\varepsilon_2\mu_2}\beta\gamma_2}K_m\left(\frac{\omega b}{v\gamma_2}\right) - \frac{1}{k_2 b}\int_{-z_0}^{z_0} dz\, e^{-i\omega z/v}\left\{\mathscr{H}^+_{\ell m}(k_2 z) + \mathscr{H}^-_{\ell m}(k_2 z)\right\}\right].
\end{aligned} \tag{F.91}$$

The factor $e^{im\phi_0}$ in Eqs. (F.90) and (F.91) cancels out the factor $e^{-im\phi_0}$ inside the expansion coefficients $a^{\text{II}}_{\ell m}/b^{\text{II}}_{\ell m}$, and can therefore be omitted in all terms.

The contribution of the inner trajectory can be further split into the contributions of the bulk modes and the Begrenzung effect $\Gamma_{\text{I}} = \Gamma_{\text{bulk}} + \Gamma_{\text{Begr}}$, where

$$\begin{aligned}
\Gamma_{\text{bulk}}(\omega) &= \frac{e}{\pi\hbar\omega}\text{Re}\int_{t_1}^{t_2} dt\, e^{-i\omega t}\mathbf{v}\cdot\left\{\mathbf{E}^{\text{I}}_e(\mathbf{r}_e,\omega)|_{\varepsilon_1,\mu_1} - \mathbf{E}^{\text{I}}_e(\mathbf{r}_e,\omega)|_{\varepsilon_2,\mu_2}\right\} \\
\Gamma_{\text{Begr}}(\omega) &= \frac{e}{\pi\hbar\omega}\text{Re}\int_{t_1}^{t_2} dt\, e^{-i\omega t}\mathbf{v}\cdot\mathbf{E}^{\text{I}}_{\text{B}}(\mathbf{r}_e,\omega)\,.
\end{aligned} \tag{F.92}$$

The bulk term corresponds to the electron traveling in the bulk of the material and does not contain any information of the boundary. It can therefore be calculated analytically, utilizing Eq. (F.56) with $d = 2z_0$. The Begrenzung term yields

$$\Gamma^{\text{M}}_{\text{Begr}}(\omega) = \frac{e}{\pi\hbar\omega}\text{Re}\sum_{\ell m} e^{im\phi_0}\frac{mb^{\text{I}}_{\ell m}}{\sqrt{\ell(\ell+1)}}\int_{-z_0}^{z_0} dz\, e^{-i\omega z/v} j_\ell(k_1 r_0) Y^m_\ell(\theta, 0) \tag{F.93}$$

and

$$\Gamma^{\text{E}}_{\text{Begr}}(\omega) = -\frac{e}{\pi\hbar\omega}\text{Re}\sum_{\ell m} e^{im\phi_0}\frac{a^{\text{I}}_{\ell m}}{\sqrt{\ell(\ell+1)}}\frac{1}{k_1 b}\int_{-z_0}^{z_0} dz\, e^{-i\omega z/v}\left\{\mathscr{J}^+_{\ell m}(k_1 z) + \mathscr{J}^-_{\ell m}(k_1 z)\right\}. \tag{F.94}$$

Overall, we may write[14]

$$\Gamma_{\mathrm{EEL}} = \Gamma_{\mathrm{surf}} + \Gamma_{\mathrm{Begr}} + \Gamma_{\mathrm{bulk}} \,, \tag{F.95}$$

with

$$\begin{aligned}\Gamma_{\mathrm{surf}}(\omega) = \frac{e}{\pi\hbar\omega}\mathrm{Re}\sum_{\ell m}\Bigg\{&\frac{K_m(\omega b/[v\gamma_2])}{ik_2\sqrt{\ell(\ell+1)}}\left[mb^{\mathrm{II}}_{\ell m}\mathscr{M}^*_{\ell m} - a^{\mathrm{II}}_{\ell m}\frac{\mathscr{N}^*_{\ell m}}{\sqrt{\varepsilon_2\mu_2}\beta\gamma_2}\right] \\ &- \int_{-z_0}^{z_0} dz\,\frac{\mathrm{e}^{-i\omega z/v}}{\sqrt{\ell(\ell+1)}}\left[mb^{\mathrm{II}}_{\ell m}h^+_\ell(k_2 r)Y^m_\ell(\theta,0) - \frac{a^{\mathrm{II}}_{\ell m}}{k_2 b}\{\mathscr{H}^+_{\ell m}(k_2 z) + \mathscr{H}^-_{\ell m}(k_2 z)\}\right]\Bigg\}, \end{aligned} \tag{F.96}$$

$$\begin{aligned}\Gamma_{\mathrm{Begr}}(\omega) = \frac{e}{\pi\hbar\omega}\mathrm{Re}\sum_{\ell m}\int_{-z_0}^{z_0} dz\frac{\mathrm{e}^{-i\omega z/v}}{\sqrt{\ell(\ell+1)}}&\Bigg[mb^{\mathrm{I}}_{\ell m}j_\ell(k_1 r)Y^m_\ell(\theta,0) \\ &- \frac{a^{\mathrm{I}}_{\ell m}}{k_1 b}\{\mathscr{J}^+_{\ell m}(k_1 z) + \mathscr{J}^-_{\ell m}(k_1 z)\}\Bigg]. \end{aligned} \tag{F.97}$$

and

$$\Gamma_{\mathrm{bulk}}(\omega) = \frac{e^2 z_0}{2\pi^2\varepsilon_0\hbar v^2}\mathrm{Im}\left\{\frac{1}{\gamma_2^2\varepsilon_2}\ln\left(\left[\frac{q_c\gamma_2 v}{\omega}\right]^2 + 1\right) - \frac{1}{\gamma_1^2\varepsilon_1}\ln\left(\left[\frac{q_c\gamma_1 v}{\omega}\right]^2 + 1\right)\right\}, \tag{F.98}$$

where $\gamma_1 = 1/\sqrt{1-\varepsilon_1\mu_1\beta^2}$ is the Lorentz factor evaluated in the bulk of the NP. Notice that the expressions above still encompass contributions of the direct electron field.[15] To remedy this feature we manually remove the aforementioned contributions by subtracting once more the corresponding terms calculated in the absence of the structure. One can incorporate this correction by modifying the expansion coefficients of Eqs. (F.84) as follows

$$a^{\mathrm{I}}_{\ell m} = T^{11}_{\mathrm{E}\ell}a^{0,\mathrm{I}}_{\ell m} + T^{21}_{\mathrm{E}\ell}a^{0,\mathrm{II}}_{\ell m} - a^{0,\mathrm{II}}_{\ell m} \tag{F.99a}$$

$$a^{\mathrm{II}}_{\ell m} = T^{12}_{\mathrm{E}\ell}a^{0,\mathrm{I}}_{\ell m} + T^{22}_{\mathrm{E}\ell}a^{0,\mathrm{II}}_{\ell m} - a^{0,\mathrm{I}}_{\ell m}|_{\varepsilon_1=\varepsilon_2;\mu_1=\mu_2} \tag{F.99b}$$

$$b^{\mathrm{I}}_{\ell m} = T^{11}_{\mathrm{M}\ell}b^{0,\mathrm{I}}_{\ell m} + T^{21}_{\mathrm{M}\ell}b^{0,\mathrm{II}}_{\ell m} - b^{0,\mathrm{II}}_{\ell m} \tag{F.99c}$$

$$b^{\mathrm{II}}_{\ell m} = T^{12}_{\mathrm{M}\ell}b^{0,\mathrm{I}}_{\ell m} + T^{22}_{\mathrm{M}\ell}b^{0,\mathrm{II}}_{\ell m} - b^{0,\mathrm{I}}_{\ell m}|_{\varepsilon_1=\varepsilon_2;\mu_1=\mu_2}\,, \tag{F.99d}$$

where the notation $a/b^{0,\mathrm{I}}_{\ell m}|_{\varepsilon_1=\varepsilon_2;\mu_1=\mu_2}$ indicates that the terms are evaluated by Eqs. (F.81) and (F.80) respectively, but using the optical parameters of the host medium (ε_2, μ_2) in place of the structure's (ε_1, μ_1).

---

[14] Notice that we renamed term Γ_{II} to Γ_{surf} for homogeneity.

[15] The contribution of the direct electron field can be seen if we disregard the structure and consider the electron moving in air ($\varepsilon_{1/2}, \mu_{1/2} = 1$), in which case we find $\Gamma_{\mathrm{EEL}} \neq 0$.

CL Probability According to Eq. (F.66), the photon emission probability $\Gamma_{\rm CL}$ is defined as

$$\Gamma_{\rm CL}(\omega) = \frac{r^2}{\pi\hbar\omega}\int d\Omega\,\mathrm{Re}\big\{\mathbf{E}_{\rm ind}(\mathbf{r},\omega)\times\mathbf{H}^*_{\rm ind}(\mathbf{r},\omega)\big\}\cdot\hat{\mathbf{r}}\,. \tag{F.100}$$

According to Eq. (F.67), this leads to

$$\Gamma_{\rm CL}(\omega) = \frac{1}{Z_2\pi\hbar\omega k_2^2}\sum_{\ell m}\Big[\big|b^{\rm II}_{\ell m}\big|^2 + \big|a^{\rm II}_{\ell m}\big|^2\Big]\,. \tag{F.101}$$

As in the EELS calculation, we remove the contribution of the direct electron field by modifying coefficients $a^{\rm II}_{\ell m}$ and $b^{\rm II}_{\ell m}$ as indicated by Eqs. (F.99).

Author's Biography

email: ellistamatopoulou@gmail.com
ORCID: 0000-0001-9121-911X
ResearchGate: P. Elli Stamatopoulou
Google Scholar: P. Elli Stamatopoulou

P. Elli Stamatopoulou received her Diploma in Physics from the National and Kapodistrian University of Athens (2017), and holds a Master's degree from the National Technical University of Athens (2020). In 2023 she graduated and received her PhD degree from the University of Southern Denmark. P. Elli Stamatopoulou was the recipient of the Zonta PhD scholarship for women in natural science and technology in September 2021. As of March 2024, she is a postdoctoral researcher in Karlsruhe Institute of Technology, Germany, working on the theory of light-matter interactions in plasmonic and dielectric nanosystems, and electron-beam spectroscopies.

The present publication is based on the author's thesis, which was submitted on October 2023 and defended on December 2023, obtaining the highest evaluation. The results presented in the thesis are based on the following publications:

A **P. E. Stamatopoulou**, W. Zhao, A. Rodríguez Echarri, N. A. Mortensen, K. Busch, C. Tserkezis, and C. Wolff, "Electron beams traversing spherical nanoparticles: analytic and numerical treatment". *Phys. Rev. Res.* 6, 013239 (2024). https://doi.org/10.1103/PhysRevResearch.6.013239

B **P. E. Stamatopoulou**, S. Droulias, G. P. Acuna, N. A. Mortensen, and C. Tserkezis, "Reconfigurable chirality with achiral excitonic materials in the strong-coupling regime", *Nanoscale* 14, 17581 (2022). https://doi.org/10.1039/D2NR05063C

C **P. E. Stamatopoulou** and C. Tserkezis, "Finite-size and quantum effects in plasmonics: manifestations and theoretical modelling", *Opt. Mater. Express* 12, 1869 (2022). https://doi.org/10.1364/OME.456407

P. E. Stamatopoulou, *Strong Light-Matter Interactions in Extreme Plasmonic and Mie-Resonant Systems*, Springer Theses, https://doi.org/10.1007/978-3-031-67907-0

D S. Fiedler, **P. E. Stamatopoulou**, A. Assadillayev, C. Wolff, H. Sugimoto, M. Fujii, N. A. Mortensen, S. Raza, and C. Tserkezis, "Disentangling cathodoluminescence spectra in nanophotonics: particle eigenmodes vs transition radiation", *Nano Lett.* 22, 2320 (2022). https://doi.org/10.1021/acs.nanolett.1c04754

E **P. E. Stamatopoulou** and C. Tserkezis, "Role of emitter position and orientation on silicon nanoparticle-enhanced fluorescence", *OSA Contin.* 4, 918 (2021). https://doi.org/10.1364/OSAC.412032

Other publications and preprints submitted during the author's PhD studies include:

F C. Tserkezis, **P. E. Stamatopoulou**, C. Wolff, and N. A. Mortensen, "Self-hybridisation between interband transitions and Mie modes in dielectric nanoparticles". *Nanophotonics* **13**, 2513 (2024). https://doi.org/10.1039/D2NR05063C

G J. I. Bundgaard, C. Nicolaisen Hansen, **P. E. Stamatopoulou**, and C. Tserkezis, "Quantum-informed plasmonics for strong coupling: the role of electron spill-out". *J. Opt. Soc. Am. B* **41**, 1144 (2024). https://doi.org/10.1364/JOSAB.512129

H A. Babaze, E. Ogando, **P. E. Stamatopoulou**, C. Tserkezis, N. A. Mortensen, J. Aizpurua, A. G. Borisov, and R. Esteban, "Quantum surface effects in the electromagnetic coupling between a quantum emitter and a plasmonic nanoantenna: time-dependent density functional theory vs. semiclassical Feibelman approach", *Opt. Express* 30, 21159 (2022). https://doi.org/10.1364/OE.456338

I **P. E. Stamatopoulou**, V. Yannopapas, N. A. Mortensen, and C. Tserkezis, "Manipulating the photonic Hall effect with hybrid Mie-exciton resonances", *Phys. Rev. B* 102, 195415 (2020). https://doi.org/10.1103/PhysRevB.102.195415

References

1. Arfken GB, Weber HJ, Harris FE (2012) Mathematical methods for physicists: a comprehensive guide, 7th edn. Academic Press
2. Olver FWJ, Daalhuis ABO, Lozier DW, Schneider BI, Boisvert RF, Clark CW, Mille BVSBR, Cohl HS, McClain MA (2023) NIST digital library of mathematical functions. Release 1.1.9 of 2023-03-15. https://dlmf.nist.gov/
3. Abramowitz M, Stegun IA (1972) Handbook of mathematical functions with formulas, graphs, and mathematical tables. National Bureau of Standards
4. Jackson JD (1998) Classical electrodynamics, 3rd edn. Wiley
5. Mie G (1908) Ann Phys 330:377. https://doi.org/10.1002/andp.19083300302
6. Müller EE (1987) J Math Phys 28:2786. https://doi.org/10.1063/1.527727
7. Tai C-T (1994) Dyadic green functions in electromagnetic theory. IEEE
8. Collin RE (1986) Electromagnetics 6:183. https://doi.org/10.1080/02726348608915211
9. Bohren CF, Huffman DR (1983) Absorption and scattering of light by small particles. Wiley

10. Stratton JA (1941) Electromagnetic theory. McGraw-Hill
11. García de Abajo FJ (2010) Rev Mod Phys 82:209. https://doi.org/10.1103/RevModPhys.82.209
12. García de Abajo FJ (1999) Phys Rev B 59:3095. https://doi.org/10.1103/PhysRevB.59.3095
13. Maciel-Escudero C, Reyes-Coronado A (2019) Wave Motion 86:137. https://doi.org/10.1016/j.wavemoti.2019.01.005
14. Winther A, Alder K (1979) Nucl Phys A 319:518. https://doi.org/10.1016/0375-9474(79)90528-1
15. Gradshteyn IS, Ryzhik IM (2007) Table of integrals, series, and products, 7th edn. Academic Press

The manufacturer's authorised representative in the EU is Springer Nature Customer Service Centre GmbH, Europaplatz 3, 69115 Heidelberg, Germany. If you have any concerns regarding our products, please contact ProductSafety@springernature.com

Printed and bound by CPI Group (UK) Ltd, Croydon, CR0 4YY
01/10/2026
02225966-0001